The Control *and* Treatment *of* Combined Sewer Overflows

The Control *and* Treatment *of* Combined Sewer Overflows

Second Edition

Edited by Peter E. Moffa

VAN NOSTRAND REINHOLD
I(T)P® A Division of International Thomson Publishing, Inc.

New York ▪ Albany ▪ Bonn ▪ Boston ▪ Detroit ▪ London ▪ Madrid ▪ Melbourne
Mexico City ▪ Paris ▪ San Francisco ▪ Singapore ▪ Tokyo ▪ Toronto

ITP® A division of International Thomson Publishing, Inc.
The ITP logo is a registered trademark under license

Printed in the United States of America

For more information, contact:

Van Nostrand Reinhold
115 Fifth Avenue
New York, NY 10003

Chapman & Hall
2-6 Boundary Row
London
SE1 8HN
United Kingdom

Thomas Nelson Australia
102 Dodds Street
South Melbourne, 3205
Victoria, Australia

Nelson Canada
1120 Birchmount Road
Scarborough, Ontario
Canada M1K 5G4

Chapman & Hall GmbH
Pappelallee 3
69469 Weinheim
Germany

International Thomson Publishing Asia
221 Henderson Road #05-10
Henderson Building
Singapore 0315

International Thomson Publishing Japan
Hirakawacho Kyowa Building, 3F
2-2-1 Hirakawacho
Chiyoda-ku, 102 Tokyo
Japan

International Thomson Editores
Seneca 53
Col. Polanco
11560 Mexico D.F. Mexico

1 2 3 4 5 6 7 8 9 10 01 00 99 98 97

Library of Congress Cataloging-in-Publishing Data

Control and treatment of combined sewer overflows / edited by Peter E. Moffa. — 2nd ed.
p. cm.
Includes bibliographical references and index.
ISBN 0-442-02405-3
1. Combined sewer overflows. 2. Urban runoff. I. Moffa, Peter E.
TD662.C66 1997
628'.21—dc21 97-23696
CIP

http://www.vnr.com
product discounts, free email newsletters, software demos, online resources
email: info@vnr.com
A service of ITP®

Contents

About the Editor

Peter E. Moffa is a principal with Moffa and Associates Consulting Engineers, a firm specializing in CSO abatement. He has more than 25 years of international experience in combined sewer overflow (CSO) abatement technology through involvement with over 30 municipalities. Presently Mr. Moffa is directing projects demonstrating vortex technologies and alternatives to chlorine disinfection and serves as a project reviewer for the Water Environment Research Foundation. In addition to this textbook, he has authored numerous papers, two EPA methodologies, and the textbook *The Control and Treatment of Industrial and Municipal Stormwater.*

About the Contributors

Stephan J. Nix is a professor of civil and environmental engineering and chair of the Department of Civil and Environmental Engineering at Northern Arizona University. He has nearly 20 years of experience in urban stormwater modeling and management and has published numerous papers and reports in this area. He was one of the principal contributors to the EPA's Storm Water Management Model and has directed a number of sewer system modeling studies.

Paul L. Freedman is president of Limno-Tech, Inc., a firm specializing in environmental engineering and assessment. He has over 25 years experience in evaluating and modeling water quality issues related to urban runoff, CSO, and stormwater involving over 300 locations in 25 states and provinces. Besides his predominant work for regulated clients, he also provides extensive training, technical support, and model development for EPA and state agencies on stormwater and water-quality issues.

John K. Marr is vice-president of Limno-Tech, Inc., with two decades of experience nationwide in CSO and stormwater issues for over two dozen cities. He has also been a recognized leader in advancing new methods for wet-weather, water-quality assessment.

Richard Field has been in charge of the U.S. EPA's National Urban Wet-Weather Flow (WWF) Management and Pollution Control R&D program and in that capacity has accumulated 25 years of experience in stormwater management and pollution control. He is the U.S. EPA's WWF expert, has been responsible for more than 300 R&D projects (many have been full-scale abatement facilities), and has published more than 400 papers, reports, book chapters, and books.

Thomas P. O'Connor is a member of the staff of the U.S. EPA's National Urban Wet-Weather Flow (WWF) Management and Pollution Control R&D program.

Preface

Since the first edition of this book, released in 1989, there has been considerable progress in the area of combined sewer overflow (CSO) abatement. The most notable development was the release of the National CSO Control Strategy published on September 8, 1989; the final CSO policy was officially signed April 8, 1994. The CSO problem was first recognized in the 1970s. Since that time a number of developments have taken place. The Environmental Protection Agency (EPA) embarked upon an ambitious R&D program; receiving-water impacts have been measured and documented, and facility plans have been completed and, in some cases, implemented. The CSO policy embodies the concerns of communities, various stakeholder groups, and regulatory agencies. The resulting guidance document allows for site-specific flexibility and relief from financial burden. Unfortunately, states and EPA regional offices have varied in their interpretation of the policy, resulting in some confusion among communities. Nevertheless, much progress has been made in completing facility plans, implementing BMPs, operating prototype facilities, and in some cases completing the actual abatement. Over a dozen communities are operating their facilities with measurable improvement to the receiving waters.

It is important to continue to abate CSO pollution because it remains the single most significant source of untreated human waste and a public health threat to the "swimmability" goal of the 1972 Clean Water Act Amendments. The watershed approach has been introduced as a means of putting CSO's, urban stormwater, and nonpoint pollution in proper perspective. For communities that are unsure of the actual benefits that may derive from the actions to be taken, it is fruitful at least to begin the reductions consistent with the EPA's "TMDL" (Total Maximum Daily Load) program.

This book is structured as a methodology starting with an explanation of the CSO policy and citing actual case histories as examples of how to address the CSO problem each step of the way.

Acknowledgments

A book of this nature results from the experiences of many and, in particular, the volunteer efforts of a dedicated few. I am grateful

To my coauthors, who took time from their active schedules to communicate their experiences.

To Dan, Howard, and John of Moffa & Associates, who would remind me of other responsibilities I had given them, when asked to focus their attention on the various chapters of the book.

To the clients I have served over the last thirty years for the opportunity to work on CSO technology.

And, finally, to my wife, Lisa, and my daughters, Michele, Lara, and Stefanie, for their constant support.

Chapter 1

The Combined Sewer Overflow Problem: An Overview

■ Peter E. Moffa, P.E.
Principal
Moffa and Associates Consulting Engineers
Syracuse, New York

Origin and Awareness

Perhaps one of the first forms of pollution abatement beyond the outside privy was the combined sewer; ironically, this structure has remained as one of the last vestiges of municipal pollution in most of our major cities. Initially, drainage systems were installed during the development of our older cities to relieve street flooding and to permit the transportation of goods and services. Backyard privies and cesspools provided the means for wastewater disposal (Figure 1–1). Eventually, odors from these septic systems became identifiable from several miles away and signaled the need to transport wastes elsewhere. Storm water drainage pipes were the most convenient means of resolving the problem. Through them, wastes could be conveyed to streams and rivers with more than enough dilution to remedy the problem for many years to come—or so it was thought—thus, the origin of combining sewage with stormwater, or the "combined sewer." The capacities of such sewers were generally based on some multiple of the dry-weather sewage rate. For the most part, the combined sewer provided only minimal capacity for stormwater volume. Combined trunk sewers were often designed to convey sizable storms, as much as a one-year frequency storm. However, interceptor sewers, initiated for the purpose of diverting wastewater from the receiving streams to larger water bodies or to a treatment plant, were commonly designed for two to four times the average dry-weather rates. Therefore, it became necessary to design structures that would allow relief of flows or discharges along the interceptor sewers when stormwater exceeded the capacity; these came to be known as "combined sewer overflows."

Figure 1–1. Backyard privy to storm drain (Syracuse, 1800s).

Most major municipal areas in the United States are served by a combination of sanitary sewers, separate storm sewers, and combined sanitary and storm sewers. Today it is estimated that there are between 15,000 and 20,000 combined sewer overflows (CSOs) currently in operation throughout the United States (Association of Metropolitan Sewerage Agencies, 1988), representing about 1,300 communities and serving a population of about 37,606,000 (EPA-430/9-78-006). Many of our cities have initiated pollution abatement programs, but the problem remains largely unsolved.

Federal funding for the purpose of controlling pollution at the municipal level existed as far back as 1948, when the first Comprehensive Water Pollution Control Act was passed (Water Pollution Control Federation, 1987). In 1962, Rachel Carson's *Silent Spring* had a nationwide impact, awakening Americans to the realities of pollution. Between 1962 and 1972, environmental policies underwent significant changes with a decided shift of program responsibility from the states to the federal government. The issue of water pollution evolved from the interest of a few to the concern of many. Professional engineers and scientists were joined by environmentalists and lawyers in the formulation of policies and programs. Concerns that previously focused on public health were broadened to include fish and other wildlife. The culmination of this period was the landmark legislation known as the Federal Water Pollution Control Act Amendments of 1972 (PL92-500). It was through this act that stormwater was first acknowledged as significant and CSOs were specifically addressed on a national scale.

Recognition of the Problem

The 1972 amendments were most notable in terms of both their national comprehensiveness and their specificity. Virtually all forms of pollution were recognized, and attempts were made to identify and prioritize their relative importance through national "needs surveys" required of all municipalities by the federal government in the early 1970s. These surveys served as the basis for planning the Federal Construction Grants Program authorized through PL92-500. One of the more significant sources of pollution identified in terms of abatement costs was that from CSOs. This resulted in a special report to Congress on CSOs (EPA-430/9-78-006), as well as in a 1978 needs survey entitled "Cost Methodology for Control of Combined Sewer Overflow and Stormwater Discharge" (EPA-430/9-79-003). The latter report estimated that the capital costs to meet the recreation-water-quality goal of swimmability, in terms of costs for the year 2000, will be $61.7 billion.

Initial estimates for CSO abatement were extremely high—over $100 billion—largely reflecting sewer separation; this served to alarm the regulatory authorities and initiate an ambitious federal research program. Since that time, the Environmental Protection Agency (EPA) and its predecessor organizations, through numerous research and demonstration (R&D) grants to municipalities, have made significant advances in the areas of receiving-water impact analysis, sewer-system characterization, control and treatment technology, and cost-effective analysis. Through its R&D of storm and combined sewer overflow, the EPA managed and coordinated over 300 such projects addressing the aforementioned topics. Much of this technology has been proposed through Section 201 of the 1972 amendments, commonly referred to as "facilities planning."

The most notable and clear-cut directive that improvements to combined sewer systems are specifically required as part of the area-wide planning process exists under Section 208 of the Clean Water Act of 1977 (PL95-217) and under the associated federal rules and regulations. The EPA's guidelines for completing this element of planning spelled out the steps to be followed:

1. Inventory existing combined sewer systems, including locations of intakes, bypasses, pipes, regulators, and outfalls.
2. Assess system performance.
3. Measure or estimate waste constituents, wasteloads, and wastewater flows.

4. Project wasteloads and waste flows.
5. Develop and evaluate alternatives.

Unfortunately, the nation's pollution abatement program has progressed more slowly than the original deadlines set through congressional legislation. The 1972 amendments provided for an interim goal of the protection and propagation of fish, shellfish, and other wildlife and for the achievement of recreational uses by July 1, 1983. This is commonly referred to as the "fishable, swimmable goal." The July 1, 1983, deadline was kept intact by the Clean Water Act of 1977. The Construction Grant Amendments of 1981 (PL97-117) empowered state governors with the ability to use 20% of the states' federal allotment of construction grant funds for the correction of CSOs, if deemed a major priority. Also, a separate material fund of $200 million annually was established for marine bays and estuaries. The 1981 Amendments extended the July 1, 1983, deadline to July 1, 1988. The Water Quality Act of 1987 (contained in PL100-4) kept the latter deadline intact despite evidence that many communities would not meet the secondary treatment requirement. In fact, the 1987 act did not address the issue of the deadline. As of this writing, where the fishable, swimmable goals have not been achieved, it is due largely to human-origin bacteria emanating from stormwater and CSOs. This book addresses these problems.

Overflow Structures

Since CSO structures date back to the first sewer systems, they have taken on many forms, often reflecting a local influence. This is particularly true of sewer regulators. Many major cities have their own unique designs. Generally, the more moving parts in a regulator, the more quickly it becomes inoperative. Unfortunately so, many of these structures over the years have become inoperative due primarily to lack of maintenance. All too often the monies for such maintenance fall prey to annual budget cuts.

Regulators were designed for flow diversion. Generally, any significant pollutant diversion was an incidental result. The more reliable pollutant-control regulators have been the more simple static regulators. Figures 1–2 and 1–3 (EPA-670/2-74-040) illustrate static and semiautomatic regulators, respectively. The leaping weir, as found in Seattle, Washington (Figure 1–2), provides the capability of diverting a variable depth of overflow to the interceptor by adjusting the receiving edge of the overflow pipe. Inasmuch as the solids and related pollutants tend to be in the lower layer of the trunk sewer flow, this simple contrivance can be an effective removal device; its effectiveness decreases as turbulence in the flow increases. The side weir flow is perhaps the most common

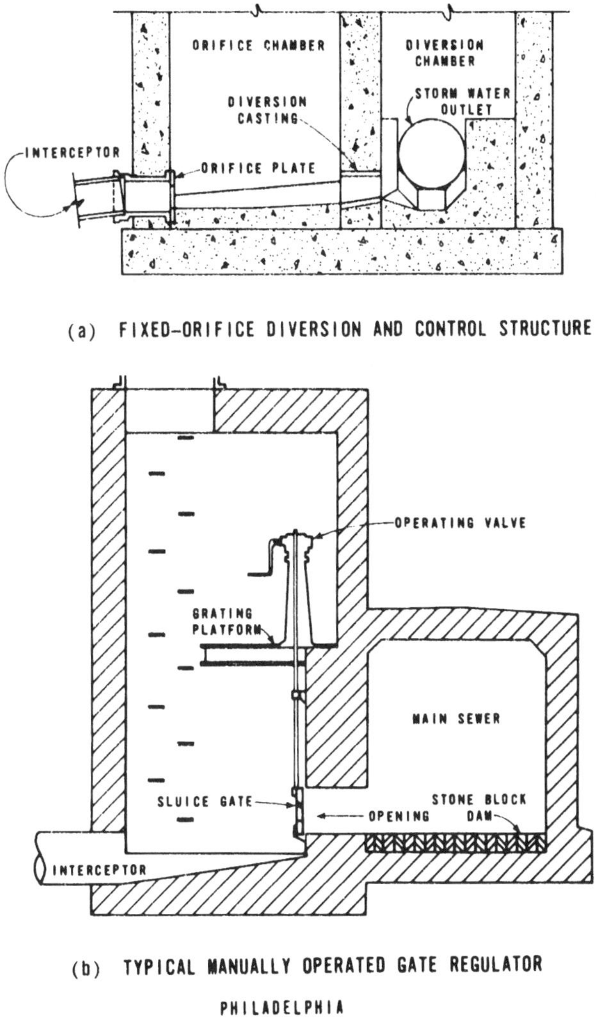

Figure 1–2. Static regulators (EPA-670/2-74-040).

overflow structure. Although the upper layer of the sewer flow tends to overflow as in the leaping weir, there is no method of controlling the quality of the diverted flow. In addition, when overflow does occur, the direction of overflow is generally perpendicular to the interceptor sewer flow, mixing solids into the upper layer, resulting in more carryover of solids than would result in the leaping weir. The controversy of complex regulators versus simple static structures will continue. However, whereas pollution concerns may force municipalities to in-

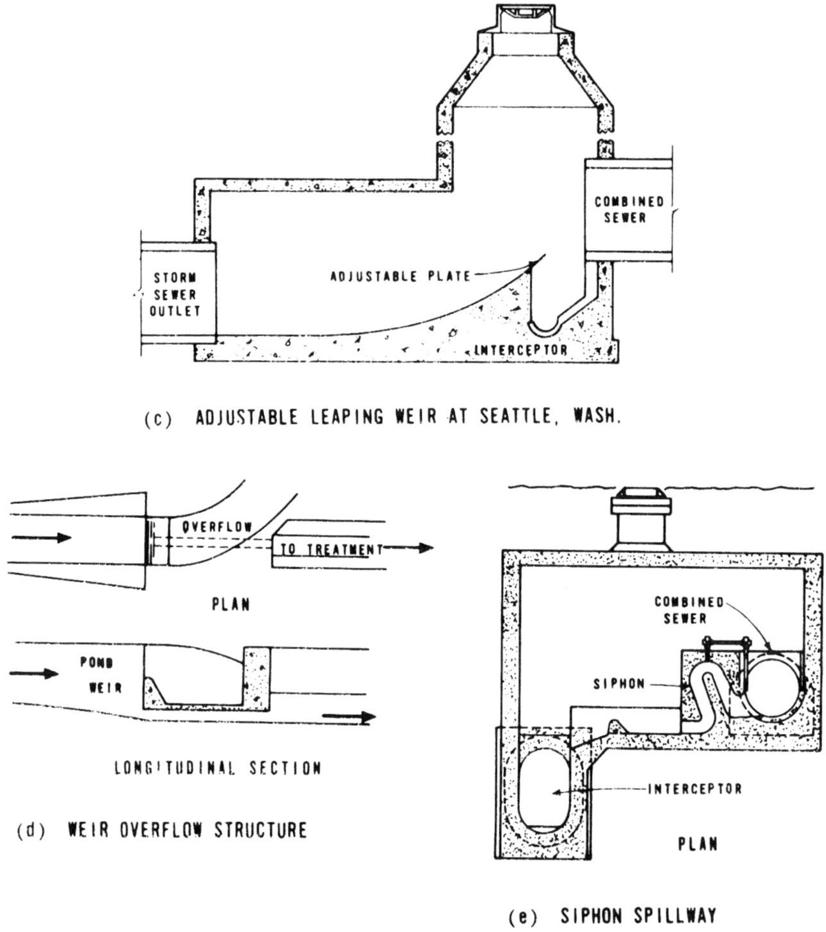

Figure 1–2. Continued.

spect and improve their sewer systems, financial constraints will continue to limit such efforts. Therefore, the guiding rule that should prevail is to keep regulators as simple as possible to ensure their operation even when ignored.

Overflow Characteristics: Quantity and Quality

The sewage component of CSOs is the smaller, and certainly the more predictable, component that can be typified by the population and industry sewered

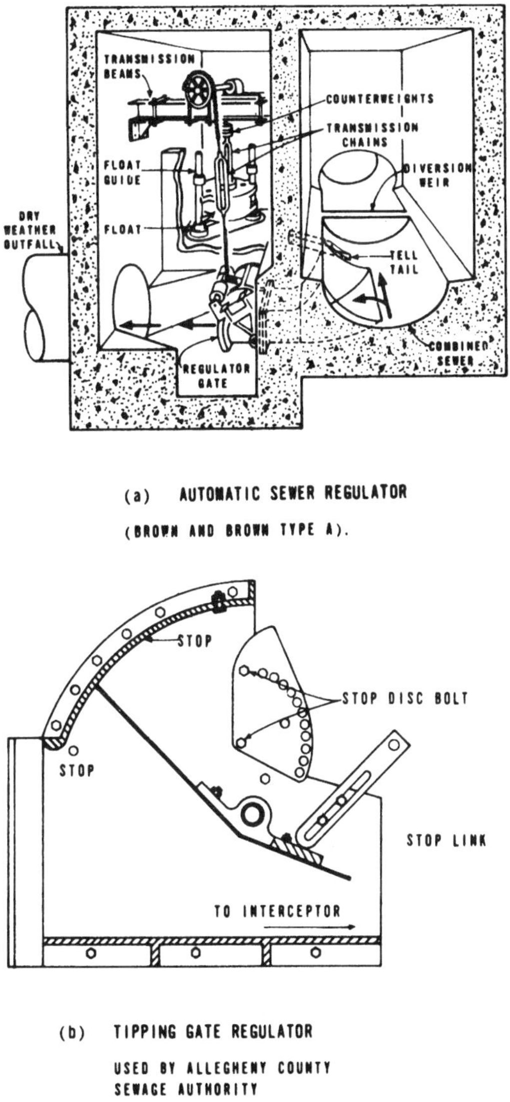

Figure 1–3. Semiautomatic regulators (EPA-670/2-74-040).

in an area. The storm-runoff component is much more difficult to characterize in that it reflects rainfall patterns, surface phenomena, and the sewer system itself. When mother nature takes a bath, so to speak, a lot goes down the drain. One thinks of a sudden rush of flow that can be observed in a drainage ditch, a trout stream, a river, or a discharge from a combined sewer system. Thus it is not surprising that the first CSO monitoring programs conducted in the early 1970s

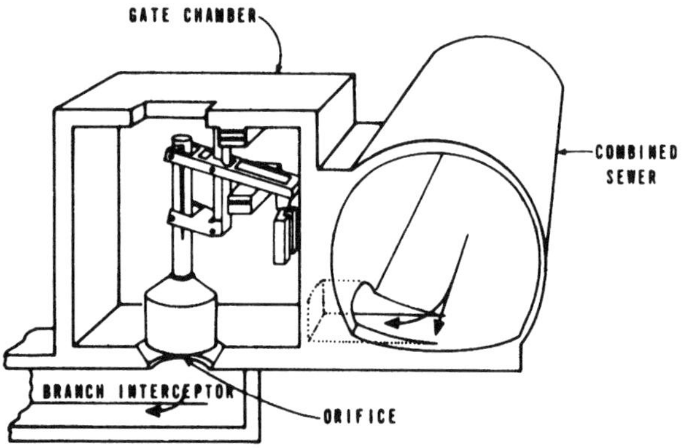

(c) CYLINDRICAL GATE REGULATOR

Figure 1–3. Continued.

identified "first-flush" effects, which can represent a quantity and quality of flow several times the average values. A typical first-flush effect is illustrated by Figure 1–4. Because of the quantity of pollutants that can discharge over this relatively short period of time, it becomes necessary to determine the quantity, evaluate the impact, and abate the pollution over this period; the peak rate of flow during this period significantly influences the basis of design and related costs of abatement.

Owing to their greater variability, achieving representative monitoring of CSOs becomes much more difficult than that of dry-weather sewage. One of the first monitoring programs was conducted in Syracuse, New York, prompted by the National Pollution Discharge Elimination System (NPDES) requirement that a permit application be filed for each CSO discharge. This occurred about the same time that needs surveys were being initiated to assess the national magnitude of the problem. At a meeting in the spring of 1973 involving Syracuse's county (Onondaga) officials, state representatives, and agents from the construction grants and discharge permit sections of the Federal Water Pollution Control Administration (a predecessor of the EPA), it was suggested that 25% of Syracuse's 84 overflows be monitored. The 25% rule was to later serve as a regulatory guide for CSO characterization. However, accumulated data from CSO programs and mathematical modeling techniques have permitted the reduction of this percentage. Table 1–1 lists CSO pollutant concentrations for several cities. Table 1–2 lists the first-flush effects normally expected from CSOs. As can be noted, the interval 0–30 minutes from the initial discharge results in noticeably higher concentrations for all parameters except fecal coliforms. Owing perhaps to the variability inherent in bacterial sampling and analysis, first-flush effects are not as evident from this parameter.

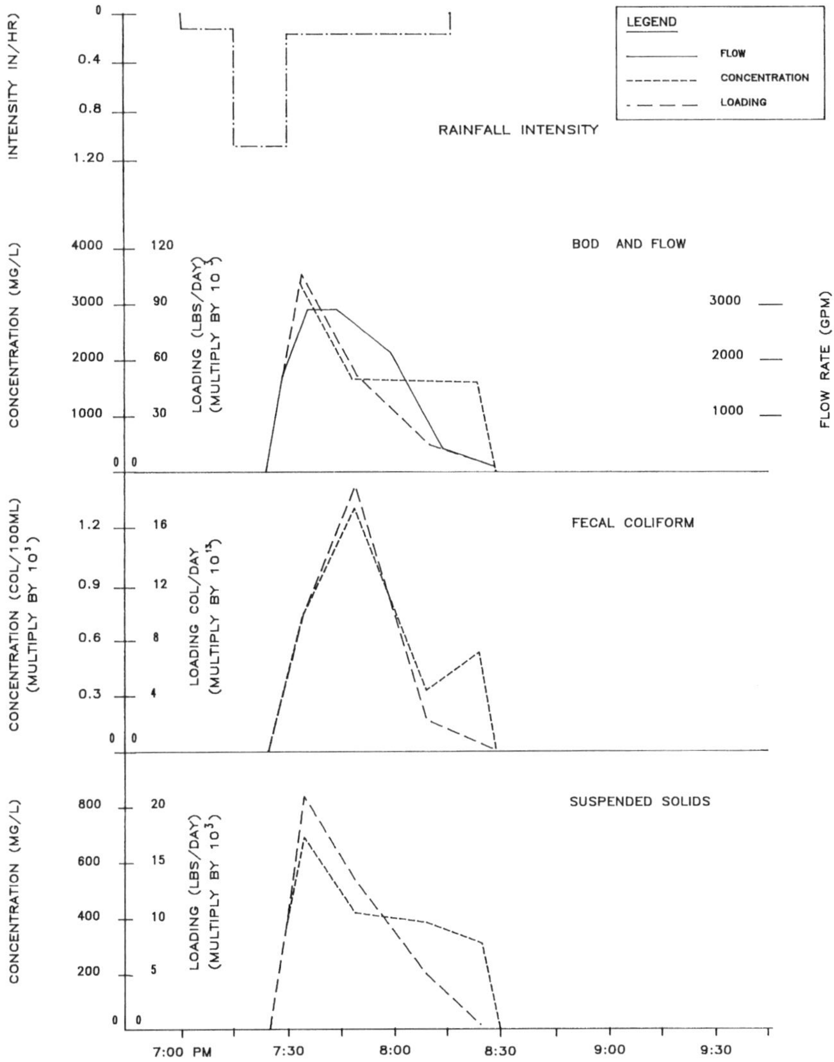

Figure 1–4. First-flush effect (Stearns and Wheler 1979).

Methods of Measurement

The demand for monitoring resulted in the development of a variety of portable and fixed-flow meters and samplers. Initial meters consisted of a float in a still well wired to a pen recorder that required manual transfer of data from a chart. More recent devices consist of ultrasonic detection that measures water level and

Table 1-1. Pollutant Concentrations in Combined Sewer Overflows

	Average Pollutant Concentration (mg/l)									
	TSS	VSS	BOD	COD	Kjeldahl nitrogen	Total nitrogen	PO_4–P	OPO_4–P	Lead	Fecal coliforms[a]
Des Moines, Iowa (11)	413	117	64	—	—	4.3	1.86	1.31	—	—
Milwaukee, Wisconsin (15)	321	109	59	264	4.9	6.3	1.23	0.86	—	—
New York City, New York										
Newtown Creek (16)	306	182	222	481	—	—	—	—	0.60	—
Spring Creek (17)	347	—	111	358	—	16.6	4.5[b]	—	—	—
Poissy, France (18)[c]	751	387	279	1005	—	43	17[b]	—	—	—
Racine, Wisconsin (19)	551	154	158	—	—	—	2.78	0.92	—	201
Rochester, New York (20)	273	—	65	—	2.6	—	—	0.88	0.14	1140
Average (not weighted)	370	140	115	367	3.8	9.1	1.95	1.00	0.37	670
Range	273–551	109–182	59–222	264–481	2.6–4.9	4.3–16.6	1.23–2.78	0.86–1.31	0.14–0.60	201–1140
Syracuse, New York[d]	306	98.4	64.3	—	2.83	—	0.35[b]	—	—	1,407
Hartford, Connecticut[e]	727	—	33	319	6.20(NH_3)	—	—	—	—	2,600

[a]1000 organisms/100 ml.
[b]Total P (not included in average)
[c]Not included in average because of high strength of municipal sewage when compared to the United States.
[d]O'Brien & Gere, 1979. "Combined Sewer Overflow Abatement Program," a report to Onondaga County, NY.
[e]Calocerinos & Spina, 1988. "Hartford Combined Sewer Overflow Facility Plan," a report to the Metropolitan District Commission, Hartford, CT.
Ref. EPA 600/8-77-014 *Urban Stormwater Management and Technology*, Update and Users Guide.

Table 1–2. Illustration of First-Flush Effects from CSO System, Syracuse, NY

	Minutes from Initial Discharge		
	0–30	30–60	greater than 60
Harbor Brook Interceptor			
TSS-mg/l	983	531	256
BOD-mg/l	175	67	43
TKN-mg/l	4.92	2.50	1.80
TIP-mg/l	0.67	0.27	0.20
fecal coliform (colonies per 100 mL)	2,034,476	1,285,563	1,873,041
Lower Main Interceptor			
TSS	634	688	399
BOD	65	36	24
TKN	1.60	1.50	0.90
TIP	0.19	0.17	0.15
fecal coliform	395,984	808,492	119,631
Upper Main Interceptor			
TSS	631	420	270
BOD	134	79	56
TKN	6.01	3.55	3.19
TIP	0.74	0.39	0.37
fecal coliform	1,655,976	1,952,966	1,922,454

Ref. O'Brien & Gere, 1979. "Combined Sewer Overflow Abatement Program," a report to Onondaga County, NY.

velocity along a pipe section. Data are logged on a cassette that can be removed or relieved by an "interrogator" for subsequent transfer to a computer. Figure 1–5 illustrates such a device. Samplers also underwent significant improvement although not as apparent. Although early samplers were of the composite type, discrete or sequential samplers were essential for the definition of first-flush effects. Gradual improvements were made in timing mechanisms and in the manner in which sampling lines were purged between each sampler. A typical sampler is illustrated by Figure 1–6.

Receiving-Water Impact

(Receiving-water impact is addressed in detail in chapter 3.) It is generally recognized that the final measure of successful abatement is the receiving-water qual-

Figure 1–5. Typical flow-metering device.

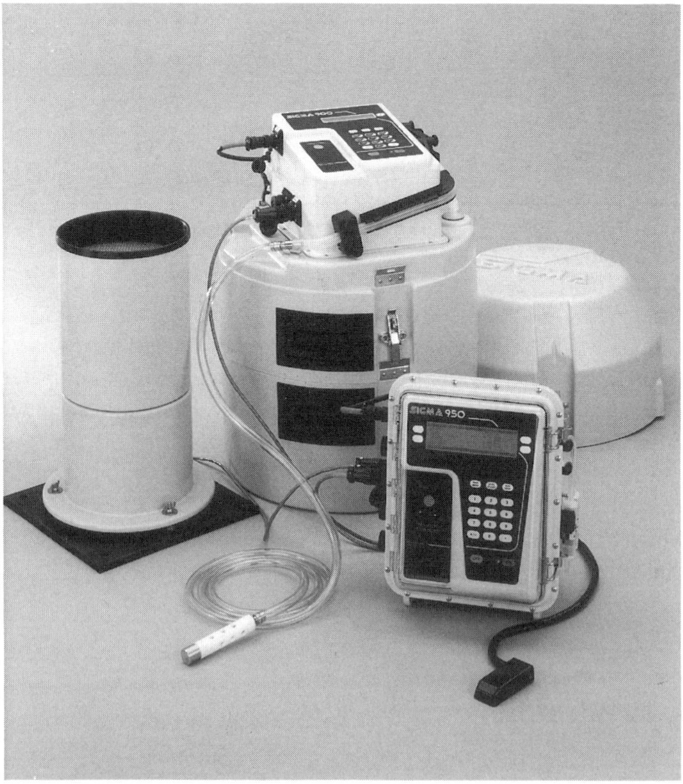

Figure 1–6. Typical sampling device.

ity. The absence of such information has led to the reliance on technology-limiting requirements, as was perhaps the case in setting the national requirement for secondary treatment of dry-weather wastewaters. However, CSO abatement defies such an approach. The storm-runoff component results in such high volumes, depending on the type of storm targeted for abatement, that a receiving-water and cost-benefit analysis becomes necessary. Regulatory agencies have generally agreed on the need for site-specific solutions. For instance, the EPA has adopted the expression "knee-of-the-curve" to determine the most cost-effective point along a cost-benefit curve. Despite this seemingly rational thinking, the EPA's enforcement division has made an interpretation that any CSO discharge must meet water-quality requirements and, as such, is a potential violation (AMSA 1988). This not only could result in prohibitively costly measures such as sewer separation but will also leave urban runoff as an unresolved problem that could otherwise be remedied through common solutions. This may require adopting water quality criteria specific to stormwater occurrences. (The current stance of the EPA and certain states is discussed later in this chapter.)

Parameters of Importance

It was not until data from several comprehensive municipal CSO studies were collected that the engineering community began to realize the types of pollutants discharged by overflow systems and their relative impacts on receiving waters. Initial concerns focused not only on the domestic pollutants associated with municipal systems but also on the variety of industrial wastes that discharge to these systems via street runoff that may contain oils, greases, heavy metals, and a wide assortment of other materials. The dilution effect of the stormwater volume was generally ignored. This effect did not become apparent until receiving-water studies were conducted on the heels of CSO system studies. Ultimately, they proved to be informative in revealing the significance of dilution during rainfall. Lack of knowledge of actual water-quality impact can result in major abatement costs disproportionate to measurable benefits. With the exception of a few isolated cases, the pollutants commonly found in municipal systems can be classified as shown in Figure 1–7 (Driscoll 1978).

Organics and Metals

The high volume of water from the stormwater component of the CSO generally provides sufficient dilution to reduce average storm concentrations of organics and metals to below levels of concern. Nevertheless, an industrialized area tributary to a sewer may cause violations; each CSO project must be evaluated inde-

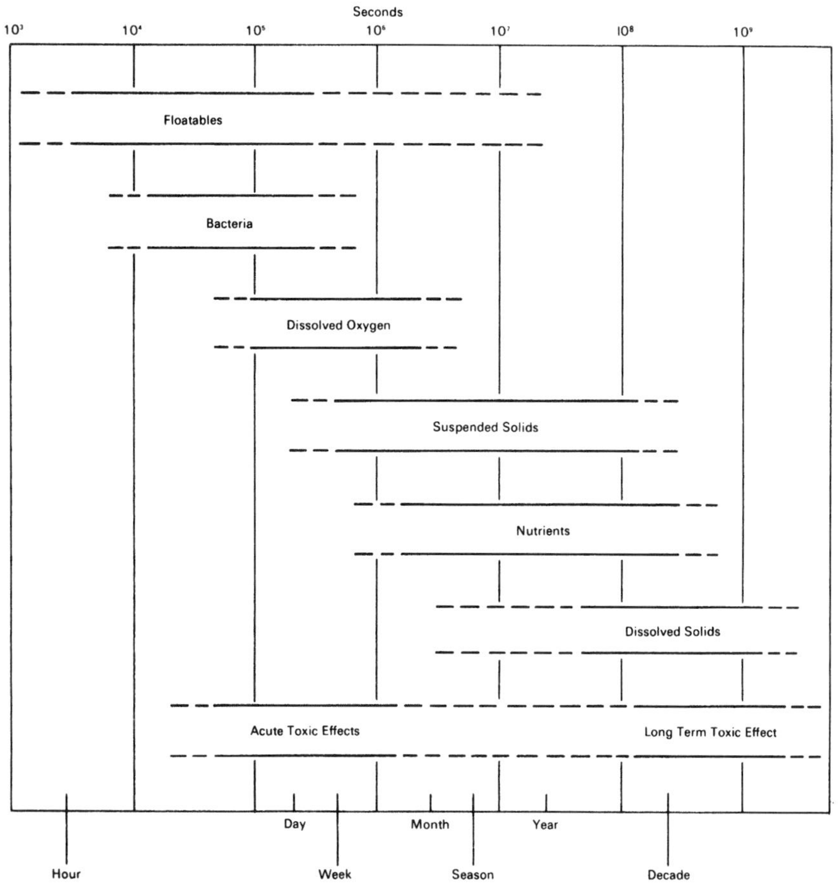

Figure 1–7. Time scales for storm-runoff-water quality problems (EPA-625/4-79-013).

pendently and with respect to the particular receiving water(s) in question. Such evaluations should be related to both acute and chronic toxicity concerns.

Dissolved Oxygen

Dissolved-oxygen impact can be classified in the form of both short-term impact largely reflecting concentrations within the overflow itself and long-term impact through downstream accumulation of BOD (biochemical oxygen demand) and subsequent resuspension. The former rarely shows a sustained measurable impact owing to dilution and time of travel, except in extremely sluggish flows and estuaries. However, the accumulative effect can provide a more complicated picture depending largely on the settling characteristics of the receiving water. Downstream accumulations can reflect more than one storm and will include upstream nonpoint sources as well.

Solids

Solids also present the two aspects of short-term and long-term impact. The former is concerned most often with aesthetics, while the latter is manifested in the form of sediment accumulation as well as oxygen demand as just described. The most obvious solids impact is that of aesthetics from floatables and debris. Such impact will play an increasing role in abatement actions as federal dollars decline; local willingness to pay for abatement will depend on the more immediate and noticeable impacts. Compared to programs designed to address long-term impact, the removal of floatables and debris may be economically more acceptable and can be implemented by simpler and more straightforward abatement techniques. On the other hand, the accumulative effect is more pervasive and difficult to identify. Intensive receiving-water investigations are normally required to confirm the significance of solids accumulation and related dissolved-oxygen impact.

Bacteria

The most common and significant impact is that from bacteria. Where CSOs occur, we can expect bacterial violations of public health standards. Despite disinfection of municipal wastewater treatment plants throughout the country, bacterial violations caused by CSOs serve as the major impediment to contact recreation in many of our major water bodies. With the advent of stormwater as a pollution concern, it will become increasingly important to adopt bacterial parameters that will distinguish human-origin sources from less harmful nonpoint sources. (The historical development of bacterial standards in this country appears in chapter 3.)

The Federal Policy

On September 8, 1989, the EPA published the draft National CSO Control Strategy, which established the foundation for states to develop their own strategies and for communities to begin to appreciate the ramifications of CSO abatement. Following extensive review by several "stakeholder" groups, the EPA published the final CSO Control Policy in April, 1994. Very few states were willing to embark upon their own requirements until the federal government issued their own policy. Some communities did proceed with CSO abatement planning and certain Best Management Practices in the early 1980s but at the risk of not complying with forthcoming requirements. Generally, communities were concerned that if they should forge ahead, they would be penalized by being required to add to what they had already done. The National Strategy served to initiate dialogue and to some extent accommodate such concerns.

From the outset, the requirements for CSO abatement provided a dilemma to the regulators, which no doubt motivated the extensive R&D undertaken by the EPA in the early 1970s. The highly variable nature of CSO volumes and pollutants dictated the need to better identify the rainfall/pollutant relationships and the actual impacts resulting from them and to develop abatement technologies specific to the demands of CSO. Thus any federal position on CSO abatement would require a degree of flexibility not heretofore required for wastewater.

The federal strategy and ultimately the policy was intended to serve only as a guideline; it does not establish or affect legal rights or obligations. The ultimate obligation is to meet the requirements of the CWA. This is normally enforceable at the state level in the form of ambient water quality criteria. However, in most cases at present, there are no state criteria for wet weather, so the federal policy serves as the fundamental guidance for meeting the CWA. On the other hand, 30 states have received approval of their CSO permitting strategies.

The CSO policy was developed as the result of input from individuals, communities, public interest groups, regulatory agencies, and professional organizations such as the Water Environment Federation, CSO Partnership, and Association of Municipal Sewerage Agencies. Consequently, it evolved into a document that accommodates the latest technological approaches to defining the problem and associated impacts, and at the same time it addresses communities both large and small in their ability to address the problem. Most important, it recognizes the site-specific nature of CSOs. At the discretion of the NPDES Authority, jurisdictions with populations under 75,000 may need to comply with only the nine minimum controls identified later in this chapter. The policy also includes guidelines on affordability that provide some boundaries on the extent of abatement expenditures a community is expected to make. (An outline of the CSO policy is provided at the end of this chapter.) The following are descriptions of the more essential features of the policy.

Four key principles of the policy ensure a cost-effective approach to meeting the objectives of the CWA:

1. Providing clear levels of controls that, if a community were to adopt, would be presumed to meet appropriate health and environmental objectives.
2. Providing sufficient flexibility particularly for financially disadvantaged communities.
3. Allowing a phased approach.
4. Review and revision as appropriate for water quality standards.

The policy requires that three fundamental categories of activity be embarked upon, namely:

1. characterize the sewer system;
2. demonstrate implementation of the Nine Minimum Controls (NMCs); and
3. develop the long-term CSO plan.

Characterize the Sewer System

The purpose of developing knowledge of how the sewer system responds to wet weather is to assist in developing both the appropriate measures to implement the NMCs and to provide the basis of the long-term CSO controls. The major elements of sewer system characterization as illustrated in Figure 1–8 are:

- rainfall records
- sewer system records
- field inspections
- monitoring at selected CSO locations
- concurrent receiving-water monitoring
- mathematical modeling

The extent of monitoring is dependent on the complexity of the specific site and affordability of the community involved. As a general rule, it is beneficial to identify the specific needs and tailor abatement requirements to those needs.

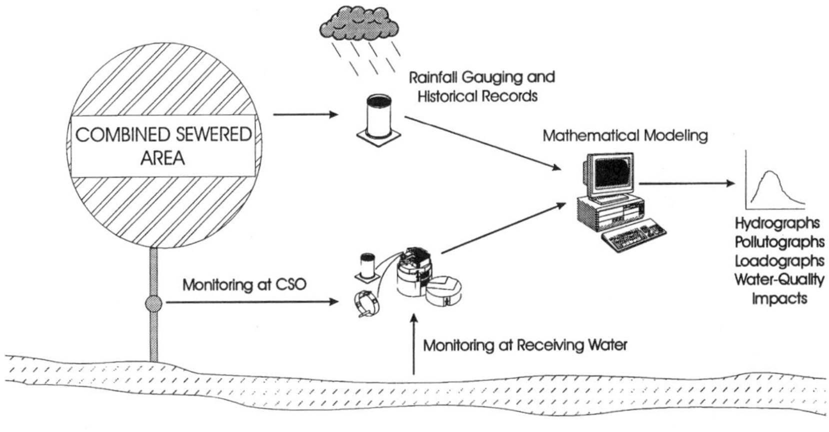

Figure 1–8. Elements of sewer system characterization.

However, the extent of monitoring needed to satisfy the scrutiny of public-interest groups and regulatory agencies can be beyond the ability of a community to pay. For this reason among others, the policy provides a series of steps that, if implemented by the community, presumes that the community complies with the requirements of the policy, hence the presumption-approach provision in the policy. This feature of the policy is of particular advantage to smaller communities; it has also proven to be of benefit to communities that undertake extensive monitoring in pursuit of demonstrating impacts and related abatement needs to allow the regulatory agencies to grant approval of a long-term plan before actual CSO impacts are demonstrated. Consensus of all parties on such impacts can take years of monitoring and in fact could continue throughout the implementation of the long-term plan. The description of the Presumption Approach and the Demonstration Approach appear later in this chapter.

Nine Minimum Controls

The purpose of the NMCs was to identify those measures that were obvious steps to take by any community not requiring extensive study. They were measures that could be considered as fundamentally "house-cleaning" procedures. They are listed as follows with brief comments where appropriate:

- Proper operation and regular maintenance programs for the sewer system and CSOs.

  Many communities are not proactive in appropriating annual budgets sufficient for the purpose.
- Maximum use of the collection system for storage.

  This measure is contingent upon the first NMC. Also, this measure requires a good knowledge of the hydraulic relationship within the system generally available only through some level of system modeling. For instance, overflow weirs can be set to maximize surcharge without causing basement flooding, but such weir settings have to be evaluated for the less frequent, high-intensity storms. Knowledge of the critical restrictive element within the combined sewer system is necessary (e.g., regulators connecting the combined trunk sewers to the interceptor).
- Review and modification of pretreatment requirements to ensure minimal CSO impacts.

  If industrial pollutants are detected in CSOs, then the first line of abatement is control at the industrial source or a more stringent pretreatment requirement.
- Maximization of flow to the POTW for treatment.

  Obviously the preferred location for treatment is the centralized location where some form of treatment already takes place. The relative cost of transmission and operational costs of satellite treatment will govern the point at which centralization may be preferred.

- Prohibition of CSOs during dry weather.

  Such discharges have nothing to do with the influence of stormwater but generally are the result of blockages that should be cleared regardless.
- Control of solid and floatable materials in CSOs.

  This measure most often will involve significant capital expenditures and may need to be part of the long-term plan. However, immediate low capital improvements (e.g., netting devices) can be applied to high-profile areas as agreed upon by the community and regulatory agency.
- Pollution prevention.

  The choice of separate sewers over combined sewers when the latter need replacement is generally preferred by state regulatory agencies. However, the eventual application of the stormwater regulations particularly in densely populated and industrial areas is giving pause for reconsideration of the benefit of combined sewers. With the possible exception of bacteria, the pollutants and related abatement technologies for CSOs and stormwater are similar, if not the same.
- Public announcement to ensure adequate notification of CSO occurrences and CSO impacts and posting of the location of CSO discharges are often required.

  A good public participation plan for the NMCs and the long-term plan is recommended.
- Monitoring to effectively characterize CSO impacts and the efficacy of CSO controls.

  This measure is recommended to enable identification of the real benefit a community derives from the NMCs and begin the process of long-term control. In a real sense this NMC begins with the first NMC in that good operation and maintenance will require the field knowledge that serves as input to a sewer system model such as the EPA's Storm Water Management Model (SWMM). The SWMM in conjunction with receiving-water investigations can then be used to identify CSO impacts for a wide range of storm intensities. These relationships are then revisited after imposing, through modeling, different CSO controls and developing cost/benefit relationships from them for investigating different long-term controls. This last NMC is essentially the first step in undertaking the long-term control plan.

Long-Term Plan

The long-term plan is to be submitted within two years of an enforcement action and can take the form of state permit, which is generally the most preferred, or a specific administrative or consent order. The long-term plan should contain certain essential elements such as:

- Levels of abatement specific to the site and water quality impacts;
- phased implementation to allow affordability;
- priorities on implementation in recognition of sensitive areas;
- inclusion of the public throughout the plan, which could span several years (e.g., 15 to 20 years);
- provision for expansion at selected sites, should greater levels of abatement be required in the future.

The features of the policy that can be considered perhaps the most helpful in assisting communities to address their CSO problem are:

- Identification of the "Presumption" and "Demonstration" approaches.
- The guidance on affordability.

Presumption Approach

This approach allows a community to provide an adequate level of control to meet the water-quality based requirements of the CWA if one of the following conditions is met:

1. No more than an average of four (4) overflow events per year. The permitting authority may allow up to two additional overflow events per year. An overflow event is one or more overflows from a combined sewer system (CSS) as a result of a precipitation event that does not receive the minimum treatment specified as:
   - Primary clarification defined as removal of floatation and settleable solids equivalent to primary treatment.
   - Disinfection of the effluent to meet water quality standards including removal of harmful disinfection chemical residuals.

Or, if the first condition is not met:

2. The elimination or the capture for treatment of no less than 85 percent by volume of the combined sewage collected on a system-wide annual average basis; or
3. The elimination of removal of no less than the mass of the pollutants, identified as causing water quality impairment through the sewer system characterization, monitoring, and modeling effort, for the volumes that would be eliminated or captured for treatment under paragraph 2 above.

The third criterion identifies the fundamental basis for the Presumption approach, which in essence is to meet the requirements of the CWA under wet-weather conditions; these requirements are more clearly identified under the Demonstration approach. The major advantage of the Presumption approach is that it allows small communities that cannot afford extensive water-quality monitoring and impact studies to proceed with abatement planning. Even in the case of larger communities, the Presumption approach provides an escape for any stalemate over water-quality impact results until a consensus is reached. Several wet-weather seasons of data may need to be collected before such consensus is reached.

The reference to primary treatment warrants clarification with respect to the type of facilities such treatment applies to. This level of treatment was intended to apply to the CSO component that is conveyed to the main sewage treatment plant (STP) and not to any satellite treatment facilities specifically designed for CSO abatement. The requirement of primary treatment is further defined as "removal of flotation and settleable solids," which presents no problem at the STP since primary settling tanks are generally capable of such treatment for the combined sewage flows that are conveyed by the interceptor. However, what level of treatment is required at other locations throughout a combined system? In other words, why should satellite locations be required to provide primary treatment? Satellite treatment often provides a more cost-effective abatement approach due to the high cost of transmission of such flows to the STP. Satellite facilities may be able to provide conventional primary treatment but will require higher capital and operation and maintenance (O&M) costs than may be justifiable on the basis of actual water-quality impacts. Treatment levels can range from simple removal of floatables to BOD capture and disinfection as illustrated by Figure 1–9.

The degree of solids removal should be based upon water-quality impact analysis, which in turn will determine the type of device (e.g., vortex) or tanks required; the type of device or tank will determine the type and quantity of solids removed. Vortex devices will remove the heavier gross solids (e.g., S.G. of 2.65 of a particle size of 0.4 to 1.0 mm and an S.G. of 1.2 at a particle size of 1.0 to 3.0 mm). Primary tanks will remove more of the lighter particles, including organics S.G. of 1.2 at a particle size of 0.03 to 1.0. This subject is covered in greater detail in chapter 4.

Demonstration Approach

The community that undertakes this approach should demonstrate each of the following:

1. The long-term plan is adequate to meet WQS and protect designated uses, unless uses cannot be met as a result of other pollution sources (e.g., natural background conditions).

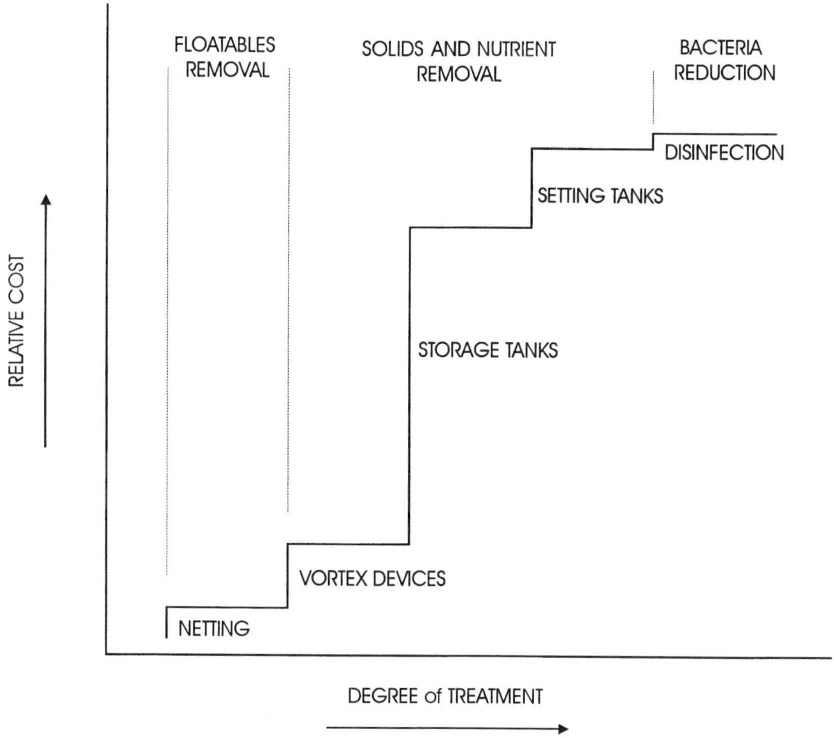

Figure 1–9. Combined sewer overflow treatment levels.

2. The CSO discharges remaining after the plan implementation will neither preclude the attainment of WQS or designated uses of the receiving waters nor contribute to their impairment. Where these conditions are not met in part because of natural background conditions or other pollutant sources than wasteload, allocations should be used to apportion loads. The total maximum daily load should serve as the basis for such allocations.
3. The long-term plan will provide the maximum pollution reduction benefits reasonably attainable.
4. The plan is designed to allow cost-effective expansion if additional controls are necessary.

An example of how the Presumption approach provided the basis for proceeding with a long-term Demonstration approach appears in chapters 2 and 5 of this book: the Syracuse, NY, case history.

Since the issuance of the policy published on April 19, 1994, a number of CSO Related Guidance Documents have been issued as follows:

- *Combined Sewer Overflows—Guidance for Long-Term Control Plan* (EPA 832-B-95-002)
- *Combined Sewer Overflows—Guidance for Nine Minimum Controls* (EPA 832-95-003)
- *Combined Sewer Overflows—Guidance for Screening and Ranking* (EPA 832-B-95-004)
- *Combined Sewer Overflows—Guidance for Monitoring and Modeling* (EPA 832-B-95-005)
- *Combined Sewer Overflows—Guidance for Financial Capability Assessment and Schedule Development* (EPA 832-B-95-006)
- *Combined Sewer Overflows—Funding Options Guidance* (EPA 832-B-95-007)
- *Combined Sewer Overflows—Guidance for Permit Writers* (EPA 832-B-95-008)
- *Combined Sewer Overflow—Questions and Answers on Water Quality Standards and the CSO Program* (EPA 832-B-95-009)
- *Performance Measures for the National CSO Control Program* (AMSA 1996)

The essential elements of the guidance documents just listed have been discussed with the exception of the *Financial Capability Assessment* and *Funding Options* documents. Performance measures are discussed in the following chapters.

Financial Capability Assessments

A community's financial capability is particularly important in the case of CSO abatement since there are no objective wet-weather standards for most states; the selection of a "design" storm has become a subjective decision based upon cost-benefit relationships. In addition, other pollutants from the watershed in question may preclude the attainment of the fishable-swimmable goals of the CWA.

The policy identifies seven factors to consider in the determination of a community's financial capability:

1. Median household income.
2. Total annual wastewater and CSO control costs per household as a percentage of median household income.
3. Overall net debt as a percentage of full market property value.
4. Property tax revenues as a percentage of full market property value.
5. Property tax collection rate.
6. Bond rating.
7. Unemployment.

The following are excerpts from an excellent summary by the CSO Partnership of the documents criteria for assessing the financial capability of a community and relating such capability to compliance schedules (*CSO Bulletin,* April 1996).

> The CSO Policy provides that implementation schedules for CSO controls may be phased, based in part on a community's financial capability. The Guidance seeks to provide criteria for assessing financial capability and to relate that capability to appropriate compliance schedules. The goal of the guidance is to "provide general boundaries to aid all parties in negotiating reasonable and effective schedules" for implementing CSO controls.
>
> The Guidance proposes a two-step approach for calculating economic capability. The first step focuses on the financial impact of current and proposed wastewater and CSO controls on individual households. The second step examines the debt, socioeconomic, and financial condition of a CSO community. The results of both steps are then combined into a "Financial Capability Matrix" which is used to assess the financial burden attributable to CSO controls and to establish what EPA views as appropriate compliance schedules.
>
> **Step One.** This step calculates a "Residential Indicator" ("RI"), which is a measure of the CSO community's average cost per household for existing and proposed wastewater treatment and CSO controls, expressed as a percentage of local median household income ("MHI"). The Guidance proposes to establish three ranges for RIs with RIs of less than one percent MHI characterized as imposing a "low" financial impact, RIs from one or two percent of MHI imposing a mid-range impact and RIs above two percent of MHI imposing a high impact.
>
> **Step Two.** The second step is to consider a community's financial Capability Indicators. EPA specifies six proposed indicators in the Guidance for evaluating community debt, socioeconomic, and financial conditions which affect a community's financial capability to implement CSO controls. The results of the six indicators are used to classify the community's financial capability as "weak," "mid-range," or "strong."
>
> The results of step one (Residential Indicator) and step two (Financial Capability Indicators) are then combined in a financial Capability Matrix which provides the basis for the permittee and NPDES permitting authority to negotiate an appropriate compliance schedule. The Financial Capability Matrix divides communities into three categories based upon their calculated burden under the Guidance (low, medium, or high), with equal weight given to the Community's RI and Financial Capability Indicators.
>
> Finally, the Guidance proposes the following general boundaries for compliance schedules based upon a community's calculated burden:

Low burden:	0–5 year compliance schedule
Mid-Range burden:	5–10 year compliance schedule
High burden:	10–15 year compliance schedule

According to the Guidance, communities in the "High burden" category may be able to negotiate compliance schedules as long as 20 years.

The lack of supporting information to explain and justify EPA's selection of the Residential Indicator categories and the Financial Capability Indicator categories will make it difficult to evaluate whether the proposed categories are appropriate for CSO communities nationwide. For example, EPA does not provide any discussion of the number of CSO communities that can be expected to fall into the three categories proposed in the Financial Capability Matrix (low, mid-range, or high burden). Also, it is unclear how EPA arrived at its proposed ranges for the compliance schedules for communities falling within the three categories of financial burden.

References: Chapter 1

Association of Metropolitan Sewerage Agencies. 1988. "Draft CSO Permitting Strategy, August 1988," Bulletin No. GB88-22.

The CSO Partnership. April 1996. "The CSO Bulletin."

Driscoll, E.D., and J.L. Mancini. 1978. "Assessment of Benefits Resulting from Control of Combined Sewer Overflows," presented at EPA Technology Transfer Seminars on Combined Sewer Overflow Assessment and Control Procedures.

EPA-430/9-78-006, "Report to Congress on the Control of Combined Sewer Overflow in the United States," CH2M Hill Southeast, Inc.

EPA-430/9-79-003, "1978 Needs Survey, Cost Methodology for Control of Combined Sewer Overflow and Stormwater Discharge."

EPA-440/5-84-002, "Ambient Water Quality Criteria for Bacteria," January 1986. U.S. EPA Report No. EPA-440/5-84-002, Washington, D.C.

EPA-625/4-79-013, "Benefit Analysis for Combined Sewer Overflow Control," Seminar Publication.

EPA-670/2-74-040, "Urban Stormwater Management and Technology, An Assessment," December 1974, Metcalf and Eddy, Inc., Palo Alto, California 94303.

EPA 1987, "EPA Region 1 Policy Statement Concerning the Relationship between the Regulation of Discharges from Combined Sewer Overflows and Water Quality Standards."

Stearns and Wheler Civil and Sanitary Engineers, Cazenovia, NY. May 1979. Combined Sewer Overflows Monitoring Program and Abatement Study for the City of Ogdensburg, NY.

Water Environment Federation. April 1994. "Combined Sewer Overflow (CSO) Control Policy Final Policy (40 CFR Part 122)."

Water Pollution Control Federation. 1987. "The Clean Water Act," No. P0070JR.

Water Quality Criteria. 1968. "National Technology Advisory Committee Report to the Federal Water Pollution Control Administration," U.S. Department of the Interior, Washington, D.C.

Chapter 2

Mathematical Modeling of the Combined Sewer System

■ Stephan J. Nix
Professor and Chair
Department of Civil and Environmental Engineering
Northern Arizona University
Flagstaff, Arizona

Introduction

Overview

A combined sewer system is a complex system subjected to the dynamic stimuli of precipitation, runoff pollution loads, and sanitary wastes. The system response is a random series of overflows to a receiving water containing various levels of pollutant loads. The interaction between the combined sewer overflow (CSO) and the equally dynamic receiving water increases the complexity. A schematic of a combined sewer system is shown in Figure 2–1.

This chapter introduces and discusses the use of deterministic mathematical models to simulate the behavior of the combined sewer systems. A well-conceived and validated model can be an extremely valuable tool for predicting the system's behavior under a variety of conditions.

By replicating the system, a model can answer questions that would be impractical to answer through field measurements and observations. Unfortunately, all models are, to some degree, flawed and must be adjusted to fit the situation at hand. This process is commonly called calibration or validation. There are several models available to simulate the behavior of combined sewer systems. None are necessarily better than others: Some just fit certain needs better than others.

Some require little more than a few readily available numbers such as annual rainfall and population density. At the other end of the spectrum are models that require extensive data concerning the hydrologic characteristics of the watershed, the hydraulics of the drainage and sewer system, and the dry- and wet-

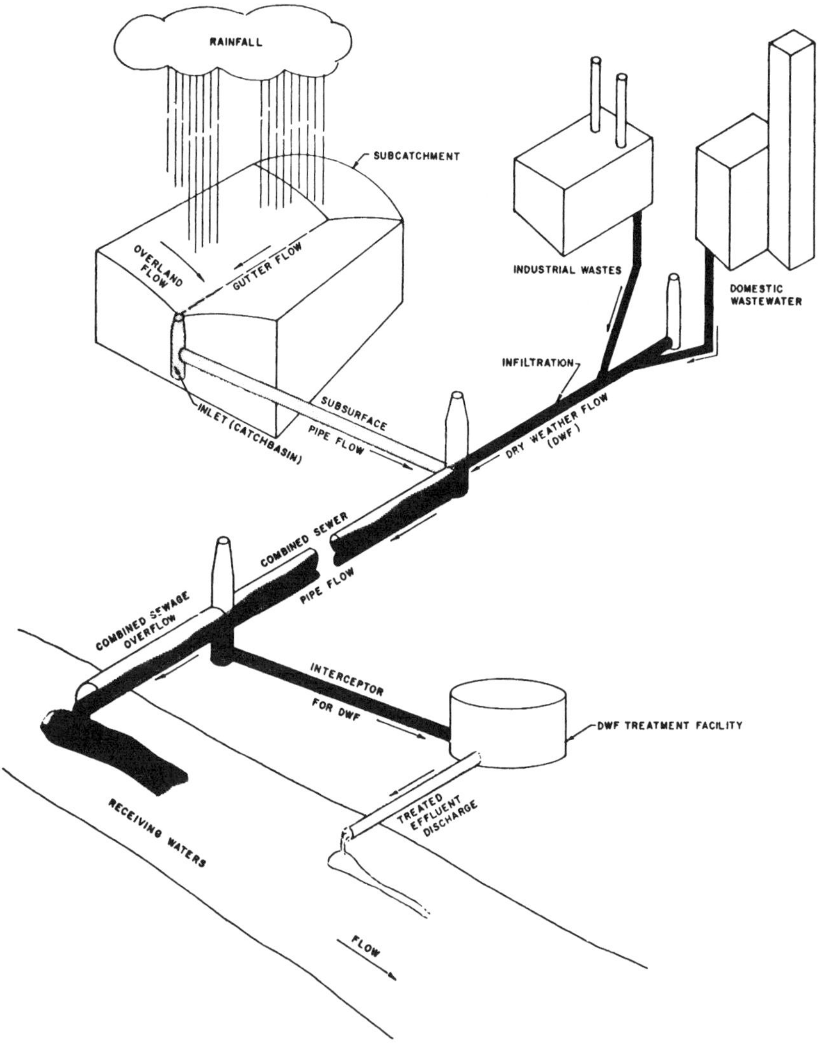

Figure 2–1. Schematic of a combined sewer system (Metcalf and Eddy, Inc., et al. 1971).

weather pollutant sources. The simpler models are more appropriate for quick, preliminary estimates of CSO volumes and pollutant loads (usually on an annual basis). The more complicated models are better suited to detailed planning and design work. Each has its place, and any study of a CSO problem may make use of more than one model.

Because all models have limitations, computer simulation can be very frustrating. The healthiest response is to recognize those limits and to realize that

CSO modeling and simulation is a science and an art. Most important, though, the user must remember that a model is just a tool and not a replacement for knowledge and judgment. The information produced by a model must be interpreted with sound understanding of the fundamental physical principles, the combined sewer system, and the model itself.

Definitions

Definitions vary in the field of hydrologic modeling. Let us establish a few so that the remainder of the chapter is clear.

system A network of interacting components capable of responding to one or more stimuli (e.g., a watershed, combined sewer system, receiving water).

model Any representation of a system by something other than the system itself.

simulation The application of a model. (The terms *modeling* and *simulation* are often interchanged, but keeping them separate is useful. An analogy with model railroading is appropriate: The construction of the scale rail system is modeling; operation of the scale model is a simulation of a rail system.)

mathematical model A model using mathematical relationships and algorithms to represent a system.

deterministic mathematical model A mathematical model designed to produce system responses or outputs (e.g., runoff) to temporal and spatial stimuli or inputs (e.g., precipitation).

stochastic mathematical model A mathematical model designed to produce a series of random responses (e.g., runoff) that have the statistical characteristics of historical responses of the system. In other words, the model does not use system inputs (e.g., precipitation) to generate the system outputs (e.g., runoff).

computer model A model in which the mathematical operations are carried out on a computer. (This facetious-sounding definition exists here to make the point that any mathematical relationship or algorithm may be a model. The rational formula is a model—placing it on a computer changes nothing except the speed at which the calculations are completed!)

digital computer model A computer model constructed and operated (ultimately) in the binary representations of a digital computer. (This distinction is probably not necessary for most, but some readers will remember the analog computers commonly used in recent decades. Digital computers have become efficient enough to render the analog computer nearly obsolete.)

variable A quantity that can vary in a particular model run.

parameter A quantity that cannot vary in a particular model run. (However, note that the definitions for variables and parameters are not as distinct as one might think. Time essentially divides the two. It is quite conceivable that a parameter [e.g., percentage of impervious area] will become a variable over time.)

algorithm A calculation procedure.

catchment The area producing the runoff passing a particular channel or stream location (e.g., gaging station).

subcatchment A portion of a catchment.

single-event simulation The use of a model to simulate the response of a catchment to a single-storm event (normally a "design" event).

continuous simulation The use of a model to simulate the response of a catchment to a series of storm events and the hydrological processes that occur between them.

run A single execution of a model.

run time The time needed to execute a model run on a computer.

simulation time The time simulated by a model run.

There are many kinds of models: Some are physical replicas, some are equations on the "back of an envelope." Some are stochastic, some are deterministic, and some are in between. Our focus here is on digital computer models that are essentially deterministic. Their aim is to simulate the response of a combined sewer system to environmental stimuli. These stimuli may be historical in nature or contrived to represent future events.

Purpose of Simulation

Why model and simulate a combined sewer system? A simple question that has a simple answer! Sonnen (1980) asserted that the intent of models is to "obviate the need for monitoring." He was overstating the case, but his point was that a well-functioning model should accurately represent the response of a system, thus reducing the need to physically monitor the system. Monitoring a combined sewer system is expensive and time consuming. Even if a system could be completely monitored, it would still not be possible to predict system behavior under changing conditions and/or the large storm not yet monitored. The purpose of simulation is to predict system responses to various stimuli and conditions.

Limitations of Simulation

The use of any model to simulate a combined sewer system and its catchment is inherently limited in a number of ways. First, a computer model cannot improve a database. It can extract information from a database, but it cannot overcome data inadequacies. Second, no model will produce completely accurate results because every model is incomplete and biased in its representation of the system. Third, numbers produced by a computer model are no more accurate than numbers produced by hand calculations, just faster. Placing a model (like the rational method) on a computer does not improve it. Computer simulation is merely a quicker way of executing a large number of calculations, all of which the user should understand and support as if he or she were doing the same calculations by hand.

Selection of a Model

Assuming that a computer model will be used to help analyze a CSO problem, the focus shifts to the selection of an appropriate model. Since it is likely that the user will have neither the time nor inclination to develop an in-house model, the chore becomes a matter of selecting one of the many off-the-shell models. No one model is perfect, and each has its advantages and liabilities. The key is first to have a firm idea of what is to be accomplished with a model and then to select the one that best fits the situation within the constraints of the available resources. This is not a trivial exercise but, at some risk, many treat it as such.

A maladapted model may well be worse than no model at all. A model that requires a large, detailed database when the project has neither the time nor money to provide the data will almost certainly result in an unsatisfactory experience. A model designed to provide rough estimates will be of little use in design work.

In theory, the selection of a CSO model should be based on careful benefit-cost analysis. Benefits and costs are often difficult to quantify, so for all practical purposes the decision is more qualitative than quantitative. Nevertheless, the basic concept is instructive. Figure 2–2 presents the basic concept of benefit-cost analysis. In general, the more complex a model, the more information it is capable of producing. (This may or may not be true, but certainly no model is intentionally given additional detail and complexity to produce less information.) However, the cost of that information is rising rapidly. What are the benefits of that information? More information is good, but as we obtain more, the next piece provides less additional value. What this means, of course, is that there exists an ideal point at which we can maximize the net gain or benefits (point A in Figure 2–2). In other words, the model that maximizes the net gain should be selected.

The content of the last paragraph represents an academic exercise. The basic message is inescapable: A model should be selected with much the same care as one would apply to any engineering decision. The decision is more complicated than the selection of a model to meet the technical goals of a particular applica-

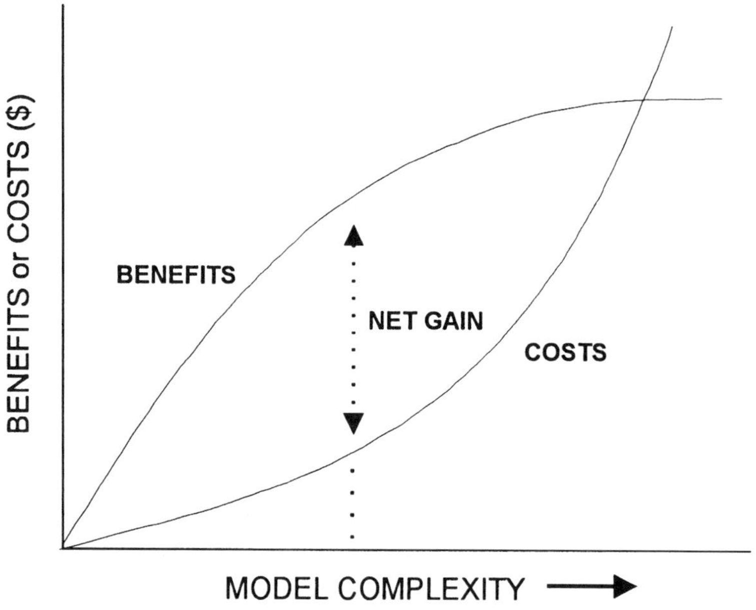

Figure 2-2. Benefit-cost analysis for model selection.

tion. In this light, it is useful to discuss briefly what other factors play an important role in such an analysis:

1. **Availability of suitable hardware.** This is not the problem it once was. Nearly all models useful for CSO work can be executed on a personal computer with the appropriate level of RAM, hard disk capacity, and so on. However, there are differing hardware requirements of which the user must be aware.

2. **Availability of trained personnel.** Water-resources models are not known for their user friendliness, and, even if they were, it takes a fairly good water-resource or environmental engineer to maintain his or her mastery over a model. In addition, the model user must be competent enough to believe his or her own judgments above and beyond the output of a model.

3. **Long-term commitment to the model.** At times it may be desirable to develop expertise with a particular model, even if that model is not ideal for the current application. If a number of similar applications are anticipated, it may be more beneficial to invest heavily in one model than to switch models from project to project.

4. **In-house model experience.** Very often, experience with a particular model is readily available. In this case, the fact that no training or warm-up period is necessary may outweigh some of the deficiencies of the model.

5. **Acceptance and support of the model.** It is quite possible that one fairly obscure model will meet the needs of a given project or series of projects. However, there is no substitute for learning from the experiences of colleagues. Thus, if the model is used by few, it becomes much more difficult to establish its credibility and to interpret its results. Similarly, if there is little or no support from the model developers (i.e., to provide updates, correct errors, provide user assistance, etc.), the chances for success are diminished. Large organizations with a substantial commitment to modeling may be able to provide such support internally, but many smaller organizations would find this difficult.

There are, to be sure, other factors to be considered. It should be clear, though, that the selection of a model is not a trivial exercise.

Catalog of Models

There are plenty of models from which to choose. Although much more should be considered, it is probably most useful to classify models for CSO work based on their routing characteristics; in other words, how much of the sewer system can be modeled? At the low end of the scale are very simple models that require a minimum of input data and often need little, if any, help from a computer. Their principal purpose is to provide the analyst with a first-cut assessment of the problem and the crudest of abatement strategies. Examples include SWMM Level I (Heaney et al. 1976; Heaney and Nix 1977) and statistically based procedures developed by Howard (1976), DiToro and Small (1979), Hydroscience, Inc. (1979), and Li and Adams (1994).

At the middle level of sophistication are models that are run on computers, generate runoff hydrographs, and allow some measure of flow routing through channels and storage basins. These are often designed to simulate the system over long periods (i.e., continuous simulation). Storage can be viewed as a basin or as storage distributed in part or all of the sewer system. An overflow can be treated as a discharge from storage. These features allow the user to estimate the number and volume of overflows from the system over lengthy planning periods. Excellent examples of this type include the Corps of Engineers' Storage, Treatment, Overflow, Runoff Model, or STORM (Hydrologic Engineering Center 1977) and the Hydrological Simulation Program-FORTRAN (HSPF, Bicknell et al. 1993).

At the high end are models capable of routing flows through a sewer system and producing hydrographs at various locations. Some are better suited for single-event simulations (e.g., because the runoff generator may not be designed to account for interevent processes), and others are capable of continuous simulation (although lengthy run times and voluminous output often make this option unattractive). As a group, these models have a wide range of specific capabilities, with some able to model almost all facets of the sewer system and others with more rudimentary features. Some of the more well-known models of this type include the EPA's Storm Water Management Model, or SWMM (Huber and Dickinson 1988), the Illinois Urban Drainage Simulator, or ILLUDAS (Terstriep and Stall 1974), and the Quantity-Quality Simulator, or QQS (Geiger and Dorsch 1980).

Why this classification scheme? Simply because the detail to which the system is described says much about the objectives of the modeling effort. If we are interested only in overflow volumes at various points in the system (and thus not in the storage characteristics of the system), then a model that routes flows through the sewers is probably unnecessary. However, if we are interested in how those overflows behave over time, the storage effect provided by the system should be analyzed.

The family of models could be classified by other means. We could divide them according to which ones are run on a computer and which are not. This is not very illuminating, however, because most are "computerized" and quite capable of running on common personal computers. In addition, knowing whether a model operates on a computer does not tell us much about how the model deals with the physical system. We could also classify models on their water-quality modeling capabilities (or lack thereof). However, those familiar with the more sophisticated models know that even their pollutant generation and routing routines (the former in particular) are only marginally useful. This is not an indictment of the modeling community; it is simply a reflection of the fact that the state of the art is not capable of reliable water-quality simulation. (This problem will be discussed in a later section of this chapter.) Suffice it to say that water-quality simulation is not a very strong suit of most models; classifying them in this manner is therefore not very useful.

A number of compendiums and comparisons of models have appeared over the last several years (American Public Works Association 1981; Brandstetter 1977; Chu and Bowers 1977; Hall 1984; Huber 1985, 1986; Huber and Heaney 1980, 1982; EPA 1983; Whipple et al. 1983). The summaries by Huber and Heaney are particularly useful. The reader should refer to these or one of the other works for a more detailed discussion of the models available for combined sewer-system modeling.

As discussed earlier, the type of model selected has much to do with the modeling objectives. Preliminary studies and assessments can probably gain some useful information with simpler, nonrouting models. Planning studies will most likely find the storage/treatment routing models of the most value. Design work can make use of the more complete routing models. However, the reader should note that the more sophisticated models are often capable of performing simpler types of analyses. In addition, some of the simpler models can guide the use of more sophisticated models by providing preliminary results. It is quite possible that a sequence of models may be appropriate. It is also conceivable that one model can be operated with various degrees of detail and complexity to serve many needs.

Model Credibility

Models are imperfect representations of the physical system. This is true because our knowledge is incomplete—our models can be no better. Therefore, it is necessary to test the algorithms for proper functioning and to adjust parameters so that the model output adequately mimics the physical system. The former is known as verification; the latter is termed validation. Many also refer to the process of validation as calibration or parameter optimization. All three terms are acceptable.

A third technique, sensitivity analysis, is useful for both model credibility and the analysis of system behavior. This technique is used to determine which model parameters are important, that is, which have a significant impact on model output. A fourth technique is the constant review of the output from the production runs (the runs of the validated model used to meet objectives of the overall study).

Verification

The task of verification is sometimes overlooked. This is unfortunate since much can be learned about model behavior with a minimal investment of effort. The verification of a model involves running a checkout of its components to determine whether they perform as expected and to learn the model's idiosyncrasies. An excellent example of a useful verification procedure is illustrated in Figure 2–3. This figure shows the well-known response of a catchment to steady rainfall. Obviously, a model that does not respond in a similar manner to a constant rainfall is not properly constructed. This is an overly simplified example, but other checks on the response of the model to controlled and well-understood stimuli can shed light on problems and build confidence in the model. These checks can also help the user understand the nuances of the model and thus make the later validation effort (as well as the production runs) more efficient.

Validation

The validation of a model requires the adjustment of model parameters so that the model accurately predicts system behavior over a wide range of conditions. The art of model validation or calibration is indeed that—an art. In spite of attempts to formalize and automate validation procedures for hydrologic models, the process is usually an iterative, heuristic exercise involving repeated model runs and subsequent adjustments. This is not meant to belittle the formalized approaches, but the state of the art is not so well developed that a competent user

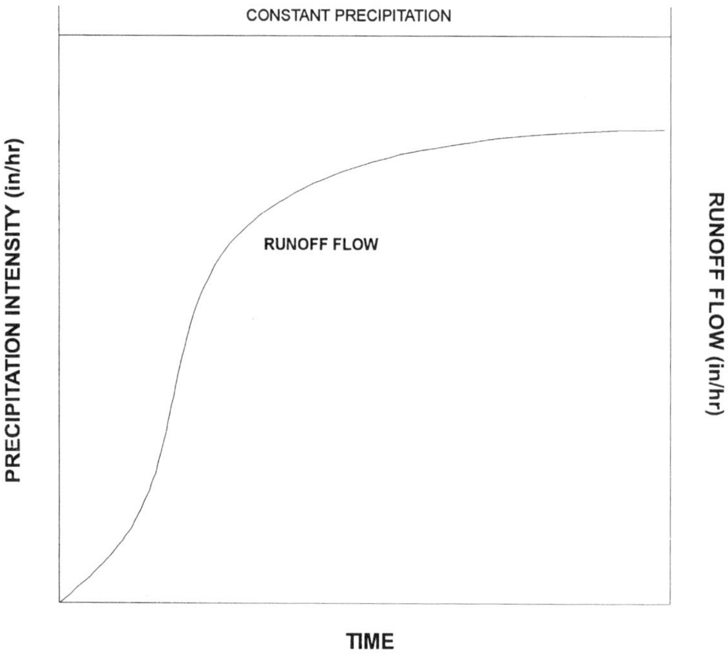

Figure 2-3. Verification exercise—response of a catchment to steady rainfall.

could not perform as well as (or even outperform) an automated calibration technique. If the user has made the investment in time to understand the model in the verification stage, adjustments can be made in an intelligent, efficient manner.

One of the most perplexing problems facing the model user is the development of the database against which the model is to be validated. How many storm events should be monitored? How many rain gages are needed, and where should they be placed? Where in the sewer system should flow and quality be monitored? No one can really provide answers to these questions. They can be answered only by a careful review of the modeling objectives and the available resources (including time, money, personnel, and previously collected data). However, it is crucial that the model user not act in a vacuum. The monitoring and modeling results should complement and guide each other. In other words, initial plans for each effort will very likely be modified as the study progresses. This is not a sign of poor planning but a realization that both efforts produce valuable insights. These efforts should be somewhat concurrent with some monitoring results available at the start of the modeling work. After the initial round, both paths should intertwine frequently.

A well-run study must be cognizant of how the results will be employed and of the likely answers it will be asked to provide. For example, if a preliminary analysis of a CSO problem strongly points to treatment devices at a few locations, then the monitoring program would certainly want to focus on those sites.

Any validation exercise should explore as wide a range of conditions as practical. In other words, calibrate the model for several storms and for several locations in the system. It is likely that a model will respond to different conditions in a somewhat inconsistent manner. Every model is incomplete in its "understanding" of the physical system. Therefore, a model calibrated for a small storm may perform poorly for a large storm and vice versa. The key is to calibrate the model over the range of pertinent conditions. For example, if the modeling effort is intended to provide information for design work, then it is probably not so important for the model to perform well for very small storms. With the previously discussed points in mind, there are some general steps that the user of any model should consider when calibrating a model to a combined sewer system and its catchment:

1. **Identify the important model algorithms and parameters.** For a given application, a particular model presents the user with several pertinent and not-so-pertinent algorithms and parameters. For example, parameters associated with a surface infiltration routine are not very important when the application is for a highly urbanized, very impervious catchment. It is important for the user to identify these algorithms and parameters so that effort and money are not wasted on them (e.g., estimating infiltration parameters). This will also simplify the validation process. Sensitivity analysis (discussed later) can help in this effort, but understanding the model and the application will be more valuable.

2. **Classify the parameters.** Model parameters range from those that are almost entirely conceptual and, as a result, very difficult to estimate (e.g., width of overland flow) to those that are very ill defined and relatively easy to measure (e.g., area). In general, it is unwise to expend a lot of effort to establish a priori estimates for the more conceptual parameters. Essentially, these parameters are the prime validation or calibration "knobs." In other words, these are the parameters that should be looked to first when adjusting the model to mimic observed data. The parameters that are more directly tied to measurable quantities should be carefully estimated before the validation process and then left alone. The more conceptual parameters should be estimated in a considered and consistent manner before model validation, but one should not waste too much time trying to "measure" them. Some help can be obtained by a review of the literature to identify applications of the model to other systems.

3. **Adjust the volume parameters.** It is probably best to first adjust those parameters most responsible for producing the overflow volume. Without a fairly accurate representation of the volume produced by a storm or storms, the rest is fairly meaningless. How does one know which are most responsible for volume?

Again, by understanding how the model represents the physical world and by sensitivity analysis.

4. **Adjust the peak-and-shape parameters.** The peak and shape of the overflow event or events should be adjusted after the simulated volumes agree reasonably well with the database. Which parameters are responsible for hydrograph peaks and shape (e.g., recession) are, again, discovered by understanding the model and by sensitivity analysis. Bear in mind, however, that the effect of changing the value of one of these parameters may alter the volume results to some degree, and other parameters may have to be readjusted in what is actually an iterative process.

5. **Adjust the water-quality parameters.** A model that is not first calibrated for quantity cannot produce reliable quality results. Therefore, parameters responsible for generating pollutant loads and their transport through the system should be adjusted after the values for the quantity parameters are felt to be adequate. Unfortunately, the state of the art in quality modeling is such that this point may be of little consequence (see later discussion).

When is a model calibrated? Again, there are no easy answers. Simply put, the process of validation or calibration is used to build confidence in the model to act as a predictive tool. A model is calibrated when it is capable of predicting at the necessary level of accuracy. There are no arbitrary standards, nor is there a need for any. The user is ultimately responsible for his or her work and thus must be satisfied that the model will provide the information needed. Although there are more sophisticated methods, most calibration efforts will be visual (e.g., by comparisons of actual and simulated hydrographs) and somewhat subjective. In other words, the user will visually inspect the results of a run for accuracy, make educated adjustments, execute another run, and end the process when satisfactory results are obtained.

Sensitivity Analysis

Sensitivity analysis is used to examine the sensitivity of the model to changes in parameters. This has at least two very beneficial uses. First, as just discussed, it is important to determine which parameters have the most effect on various aspects of the model output so that the user knows which ones to adjust in the calibration process. Second, after the model has been calibrated and is being used to make predictive estimates, it is helpful to be able to examine the effects of changes in parameter values, especially those that were used in the calibration process. Why? Because these were the parameters about which the most uncertainty exists. Therefore, it is important to establish an expected response range by varying the parameter within a range of error and to document the effect on the output (e.g., overflow volumes). This will help to quantify the uncertainty associated with the parameters.

Figures 2–4 and 2–5 each show a very effective means of studying the sensitivity of a model to different parameters in terms of different output aspects. In these particular examples, changes in the values for several parameters are depicted by a percent change (on the abscissa) from the calibrated values, and the effects are shown as percent changes from those obtained by the calibrated model. Figure 2–4 shows the technique applied to runoff volume, and Figure 2–5 demonstrates the same for the peak runoff rate. These figures explicitly illustrate the parameters that have the most impact on model output.

Monitoring the Production Runs

The credibility of a model can be enhanced by continually monitoring the output of the production runs. These runs may produce suspicious results that suggest the need for a reevaluation of the model parameters. In other words, the validation process is never truly finished.

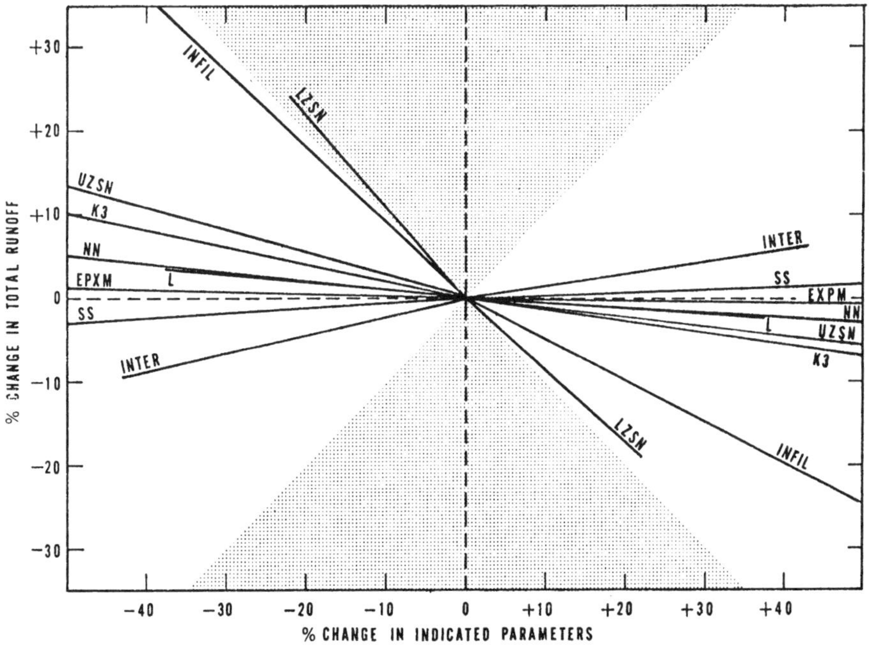

Figure 2–4. Parameter sensitivity—total runoff volume (Donigian and Crawford 1976).

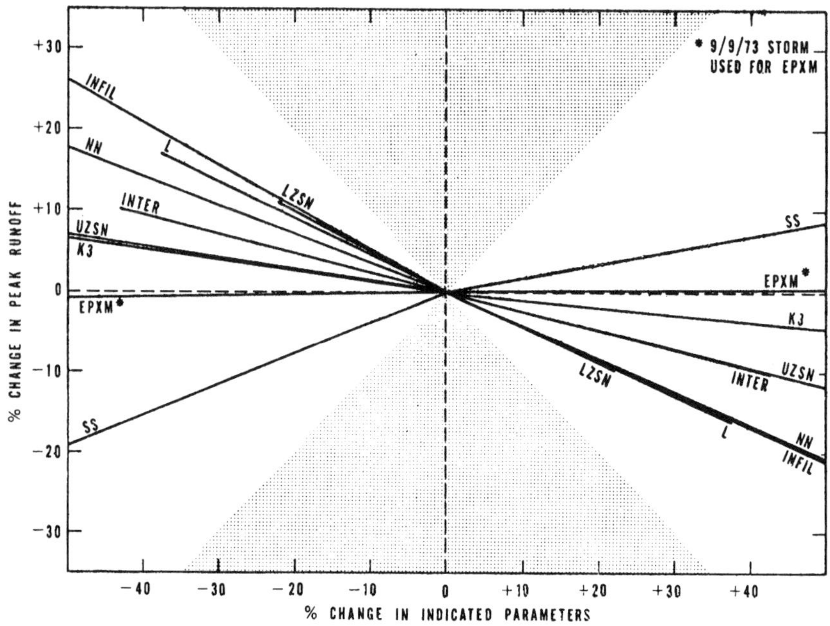

Figure 2-5. Parameter sensitivity—peak runoff rate (Donigian and Crawford 1976).

Practical Considerations

Aside from the issues raised in the previous discussions, there are a number of other practical questions that every model user should consider. These involve the basic conditions under which the model will be defined and run.

Single-Event, Multiple-Event, and Continuous Simulation

The question of whether a modeling study should focus on a single design event, several discrete design events, or a continuous series of events cannot be answered here. This is an issue that must be addressed in the context of the application. Nevertheless, there are some observations that may be useful:

1. Parameters adjusted through a calibration exercise for a single-event simulation may not be applicable to a continuous simulation for the same catchment. The effect of certain parameters on model output can be different from storm to storm.

2. The use of a "design" rainstorm (e.g., a 10-year storm) is fundamentally flawed because the resulting runoff event does not necessarily have the same re-

turn period. This is principally due to the fact that antecedent conditions have a great deal of influence on the severity of a particular storm. For example, a 2-year rainstorm may result in a 5- or 10-year runoff event if the ground is already near saturation. Continuous simulation can produce the response of a catchment to numerous years of rainfall conditions. From these responses, the user can identify critical runoff events that may be suitable for design storm simulations.

3. Continuous simulation lessens the importance of estimating the condition catchment at the start of the simulation. In single-event simulation, the initial condition of the catchment (soil moisture, water in depression storage, etc.) can have considerable effect on the model output (and, of course, the actual catchment). A long, continuous simulation (for example, a year) would be affected only for the first few events, and the overall impact would probably be minimal.

Temporal Discretization

Computer models rely on numerical approximations to solve one or more differential and/or integral equations. For example, refer to the simple continuity of mass equation:

$$\frac{dM}{dt} = m_{\text{in}} - m_{\text{out}}$$

where M is the mass stored in the system or subsystem, t is time, m_{in} is the mass inflow rate, and m_{out} is the mass outflow rate. A numerical approximation of this differential equation would involve replacing dM/dt with $\Delta M/\Delta t$. The term Δt is what is known as the "time-step" or "integration step." Some models fix the time-step, whereas others allow the user to select it. In the latter case, the user should follow the guidelines provided by the model developers, but it is helpful to investigate the effect of different time-steps, much as with any other model parameter.

In general, the time-step value has an effect on two things: (1) the ability of the model to respond to rapidly changing conditions, and (2) the stability of the numerical solution procedures. The former is demonstrated in Figure 2–6. The hydrograph labeled "A" is rapidly changing; in contrast, the hydrograph labeled "B" is much less dynamic. Hydrograph A will never be reproduced by a model using the long time-step $\Delta t'$, whereas the relatively short time-step $\Delta t''$ will allow the numerical procedures to respond to the rapidly fluctuating conditions. However, the longer time-step may well be adequate for hydrograph B and the shorter one simply unnecessary.

The stability of a model is sometimes adversely affected by a time-step that is too large. A large time-step will often cause a numerical procedure to grossly overshoot or undershoot the true solution because of the reliance of many numer-

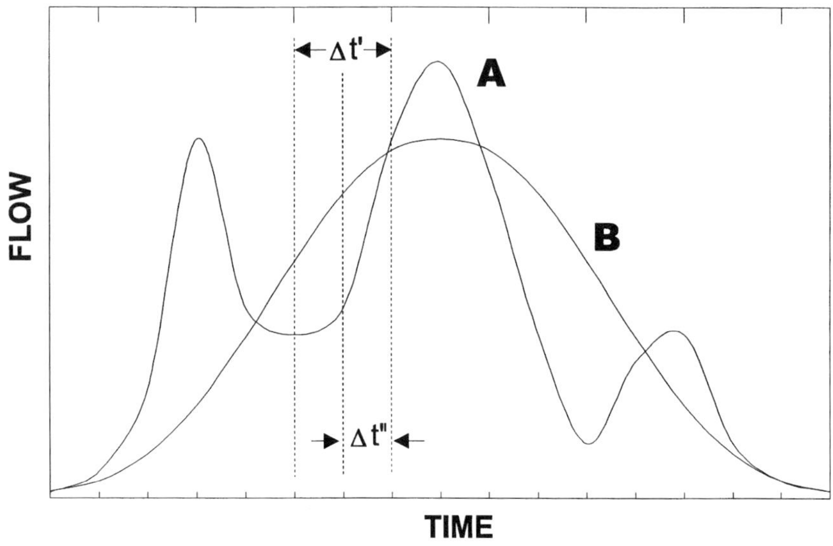

Figure 2-6. Selection of an appropriate time-step.

ical procedures on recent information (e.g., from the last time-step) to project the output value in the next time-step. If the time-step is too large, rapid changes in input conditions will not be properly reflected in the output, and the numerical procedure will probably become unstable. Since the time-step is one of the principal determinants of model run times, it is advantageous to keep the time-step as large as possible within the limits discussed previously. The optimal size will require some investigation.

Spatial Discretization

Nearly all models for combined sewer system simulation are essentially "lumped" models. This means that changes in the physical characteristics of the catchment are accounted for by assuming that they are constant over smaller portions of the catchment. This is in contrast to a model that includes the physical heterogeneities (e.g., slope) in the governing differential equations. These are called "distributed" models. Much easier to construct, a lumped model can be made to act more like the distributed model if the catchment is divided into many subcatchments and more of the sewer system is included in the model. Of course, increased system detail, or discretization, requires more effort. In general, one can say that the more heterogeneous the catchment and sewer system, the more detail one should probably include in the model. However, additional detail that adds nothing in terms of results is wasteful. As one might suspect, the level of discretization used is strongly tied to the modeling objectives and how the model responds to different levels of discretization.

An example is useful. Figure 2–7 demonstrates the response of the USEPA Storm Water Management Model (see later discussion) to various discretization schemes for a catchment in Baltimore, Maryland. One can see that the runoff hydrographs produced by the 12-subcatchment and 5-subcatchment schemes are not very different. In this case, little seems to be gained by breaking the catchment into 12 subcatchments. Would fewer than 5 be adequate? Perhaps. It is important that enough detail be left to allow all modeling objectives to be satisfied. For example, if changes in a number of smaller trunk sewers are anticipated, then it would behoove the user to include that part of the sewer system in the model. Similarly, if heavy development is predicted for a particular area of the catchment, it would be wise to include that area as a separate subcatchment so that the changes in model parameters are more straightforward.

Quality Simulation

Several years ago, Sonnen (1980) claimed that the state of the art in modeling urban stormwater quality was fairly dismal. Little has changed. This is not neces-

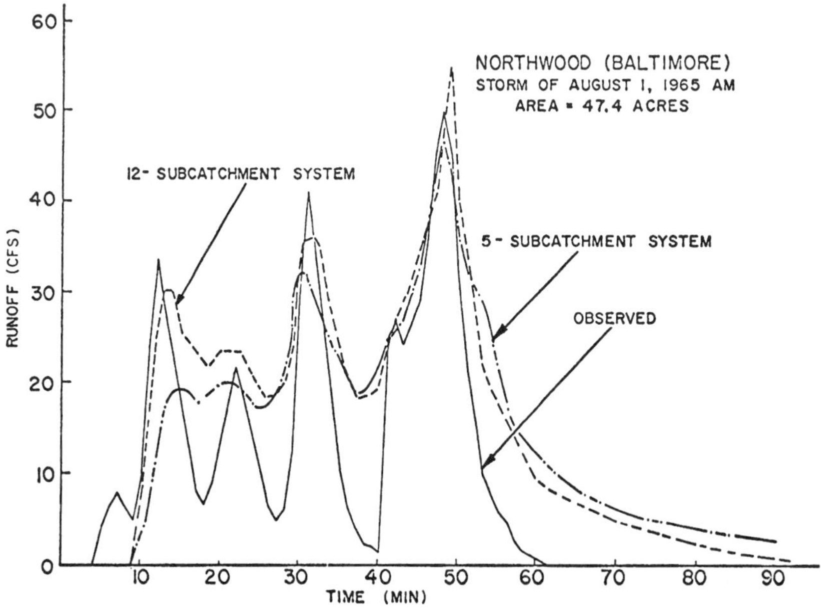

Figure 2–7. Effect of different discretization levels—Northwood catchment, Baltimore, Md. (Metcalf and Eddy, Inc., et al. 1971).

sarily the fault of those in the field. It is much more likely that the physical processes are so complex as to defy efforts to reduce them to mathematical statements. Although many modeling studies can point to a successfully calibrated quantity component, fewer can point to comparable success on the quality side. Cases in which the actual and simulated water-quality data differ by several times are not uncommon. Most model parameters and algorithms used to generate pollutant loads are extremely conceptual and thus are difficult to link with measurable physical parameters. In addition, the number of model parameters that must be estimated in order to produce water-quality output is often substantially more than that needed for the hydrologic and hydraulic output. Are the results worthy of the effort? Perhaps not.

If CSO problems are water-quality problems, can a computer model be of much help? Absolutely! Even though the problem is one of water quality, most of the currently accepted abatement technologies are principally designed by quantity considerations. There is enough experience with many of these technologies to know approximately how well they will reduce pollutant loads under various hydraulic conditions. If this is not satisfactory, pilot studies can be conducted.

This does not mean that pollutant loads cannot or should not be estimated. But at this time, a more reliable way may be to analyze the influent records at the wastewater treatment plant receiving the intercepted flows in order to estimate pollutant concentrations during wet-weather events (Mueller and DiToro 1983). Sequential sampling can be used to analyze variations in concentrations (e.g., during the first flush). Furthermore, specific CSOs can be sampled to measure local conditions. These concentrations can be plotted over time to yield a pollutograph, which can then be combined with a storm-specific hydrograph, developed through modeling to yield a loadograph:

$$\text{pollutograph} \times \text{hydrograph} = \text{loadograph}$$

This technique is not a panacea (e.g., how might the pollutant loads change when conditions in the catchment change?), but it is probably preferable to the estimates provided by current modeling techniques. Together with the overflow volume and flow information provided by a calibrated computer model, these concentrations can produce reasonable estimates of pollutant discharges by CSOs.

Another alternative is for the user to develop a simple water-quality algorithm to add to the existing computer model. This offers at least two strong advantages. One, the user will understand the algorithm (and presumably be better able to document and defend it). Second, the user can tailor the algorithm to the situation at hand. Models intended for a variety of users in a number of different applications have to cover a lot of ground. But if, for example, monitoring data indicate a strong relationship between concentration and the overflow discharge rate, then this can be simply and easily included.

The EPA Storm Water Management Model

Overview

The EPA Storm Water Management Model (SWMM) is a large, complex model capable of simulating the movement of precipitation and pollutants from the ground surface through pipe and channel networks, storage/treatment units, and finally to receiving waters. The model may be run for a single event or on a continuous basis for extended periods of simulation time.

SWMM has been released under several different official versions (Metcalf and Eddy, Inc., et al. 1971; Huber et al. 1975, 1984; Roesner et al. 1984; Huber and Dickinson 1988; Roesner et al. 1988). The official versions were primarily designed for mainframe use, but the later versions have been modified for use on personal computers. The current "official" version, version 4, is available on disks or through the Internet from the EPA at the following addresses:

Center for Exposure Assessment Modeling (CEAM)
U.S. Environmental Protection Agency
960 College Station Road, Athens, Georgia 30613
Phone: 706-355-8400
E-mail: ceam@epamail.epa.gov
WWW: ftp://ftp.epa.gov/epa_ceam/wwwhtml/ceamhome.htm

The complete model documentation (Huber and Dickinson 1988; Roesner et al. 1988) can be obtained from the National Technical Information Services (NTIS) or Oregon State University:

National Technical Information Service
5285 Port Royal Road
Springfield, Virginia 22161
Phone: 703-487-4650
Note: Huber and Dickinson (1988) is NTIS no. PB88 236 641 and Roesner et al. (1988) is NTIS no. PB88 236 658.

Dr. Wayne C. Huber
Department of Civil Engineering
Oregon State University
Corvallis, Oregon 97331
Phone: 503-737-4934
E-mail: huberw@ccmail.orst.edu

Proprietary versions of SWMM are also available. No comments or endorsements will be made here. Information is readily available through advertisements and the Internet.

There are other models capable of simulating the urban runoff process and combined sewer systems. Some probably do certain tasks better than SWMM. However, SWMM has retained its popularity over the years because of its overall performance and the considerable body of experience that new users can draw upon. No other model has seen as much discussion in the literature or as many practical uses. Its longevity and productivity have been impressive.

Useful reviews of SWMM-related literature have been prepared by Huber et al. (1985) and James (1993). Interested readers are encouraged to consult these documents.

SWMM is divided into several blocks, each designed to handle a separate phase of the runoff process or, if desired, operate independently. Some of these blocks (i.e., RUNOFF, TRANSPORT, EXTRAN, and STORAGE/TREATMENT) are computational blocks responsible for the hydrologic, pollutant generation and transport, and hydraulic calculations. Others (i.e., EXECUTIVE, GRAPH, COMBINE, RAIN, TEMP, and STATISTICS) perform various service functions. A general operational schematic of SWMM is shown in Figure 2–8, and an overview is presented in Table 2–1. The tabular summary is adapted from the user's manual for the latest version of SWMM (Huber and Dickinson 1988); it, in turn, has been adapted from material appearing in publications by Huber and Heaney (1980, 1982). Beginning with the computational blocks, the fundamentals of each are briefly described in the following text; the user's manual should be consulted for more details.

RUNOFF Block

The RUNOFF Block generates surface runoff and pollutant loads in response to precipitation and surface pollutant accumulations. In many ways, RUNOFF is the "heart" of SWMM. The manner in which surface runoff is generated by RUNOFF is actually very simple and is summarized in Figure 2–9. Essentially, a subcatchment is treated as a nonlinear reservoir that has a single inflow: precipitation. There are several outflows including infiltration, evaporation, and surface runoff. Infiltration occurs only if the ground surface is considered pervious (as opposed to an impervious surface, like a paved parking lot, which by definition allows no infiltration). The infiltration process is governed either by Horton's equation or the Green-Ampt equation with the various parameters provided by the user. Infiltrated water is routed through upper and lower subsurface zones and may contribute to total runoff through groundwater flow.

Evaporation rates are provided directly by the user and are used to lower the water in the reservoir (if there is any) and to estimate evapotranspiration from the subsurface zones. Surface runoff occurs only when the depth of water in the reservoir exceeds the depth of depression storage, which refers to surface ponding, surface wetting, interception, and other storages that must be satisfied before

Table 2-1. Characteristics of the EPA Storm Water Management Model (Huber and Dickinson 1988)

Applicable Land Drainage Area
(1) Urban **(2)** General nonurban

Time Properties

(1) Single-event or continuous simulation; both modes have an unlimited number of time-steps. **(2)** Precipitation: input at arbitrary time intervals for single-event simulation (typically 1–15 min) and continuous simulation (typically 1 h); for snowmelt daily max-min temperatures required for continuous simulation, temperatures at arbitrary intervals for single events. **(3)** Output at time-step intervals (or multiples); daily, monthly, annual, and total summaries for continuous simulation. **(4)** Time-step arbitrary for single-event (typically 5 min) and continuous (typically 1 h) simulations; variable time-step available in RUNOFF Block; time-step for Extended Transport Block (EXTRAN) routing depends on stability criteria, may be as small as a few seconds.

Space Properties

(1) Small to large multiple catchments. **(2)** Surface: lumped simulation of surface flow with allowance for up to 250 subcatchments and 10 input hyetographs, up to 250 channel/pipes may be simulated by nonlinear reservoir routing. **(3)** Channel/pipes: one-dimensional network, up to 500 conduit/nonconduit elements for TRANSPORT Block, up to 500 conduits in EXTRAN Block. **(4)** Catchment area may be disaggregated and modeled sequentially for simulation of areas too large for existing SWMM dimensions. **(5)** Storage/treatment simulated separately, receiving input from upstream routing. **(6)** Output from surface, channel/pipe, or storage/treatment simulation may serve as new input for further simulation by same or different blocks.

Physical Processes

(1) Flow derived from precipitation and/or snowmelt; snow accumulation and melt simulated using temperature-index methods developed by National Weather Service; snow redistribution (e.g., plowing, removal) may be simulated. **(2)** Overland flow by nonlinear reservoir using Manning's equation and lumped continuity, depression storage, integrated Horton or Green-Ampt infiltration (with optional subsurface routing), recovery of depression storage via evaporation between storms during continuous simulation, also exponential recovery of infiltration capacity. **(3)** Subsurface routing only of flows through unsaturated and saturated zones simulated using lumped storages; subsurface outflow by power equation; simulation of evapotranspiration and water-table fluctuation. **(4)** Channel/pipes:

(continued)

Table 2-1. Continued

(a) nonlinear reservoir formulation for channel pipes in RUNOFF Block, includes translation and attenuation effects; (b) modified kinematic wave formulation in original TRANSPORT Block assumes cascade of conduits, cannot simulate backwater over more than one conduit length, surcharging handled by storing water at surcharged junction pending available flow capacity; (c) EXTRAN Block solves complete Saint-Venant equations including effects of backwater, flow reversal, surcharging, looped connection, pressure flow; (d) infiltration and dry-weather flow may enter conduit of either transport simulation. **(5)** Storage routing using modified Puls method assuming horizontal water surface; outlets include pumps, weirs, orifices. **(6)** Surface quality on basis of linear or nonlinear buildup of dust/dirt or other constituents during dry weather and associated pollutant fractions, power-exponential washoff with decay parameter a power function of low rate only (rating curve): erosion by Universal Soil Loss Equation. **(7)** Dry-weather flow quantity and quality on basis of diurnal and daily variation, population density, and other demographic parameters, buildup of suspended solids in conduits by dry-weather deposition using Shield's criterion. **(8)** Quality routing by advection and mixing in conduits and by plug flow or complete mixing in storage units, scour, and deposition of suspended solids in conduits (in TRANSPORT Block) using Shield's criterion. **(9)** Storage/treatment facility simulated as series-parallel network of units, each with optional storage routing. **(10)** Treatment simulation: (a) use of arbitrary user-supplied removal equations (e.g., removal as exponential function of residence time); (b) use of sedimentation theory coupled with particle-size-specific gravity distribution for constituents.

Chemical Processes

(1) Ten arbitrary conservative constituents in RUNOFF Block, rainfall quality included, choice of concentration units is arbitrary; erosion of sediment is optional. **(2)** Four constituents may be routed through the original TRANSPORT Block (with optional first-order decay), three through the STORAGE/TREATMENT Block, and none through EXTRAN (quantity only).

Biological Processes

(1) Coliform washoff may be included. **(2)** Biological treatment may be simulated.

Economic Analysis

Amortized capital plus operation and maintenance costs for control units are determined.

Mathematical Properties

(1) Physically-based model. **(2)** Surface quantity: iterative solution of coupled continuity and Manning equations, Green-Ampt or integrated form of Horton infiltration (infiltration rate proportional to cumulative infiltration, not time). **(3)** Surface channel/pipe routing: nonlinear reservoir assuming water surface parallel to invert. **(4)** Channel/pipes: (a) TRANSPORT: implicit finite difference solution to modified kinematic wave equation; (b) EXTRAN: explicit finite difference solution to complete Saint-Venant equations, stability may require short time-step. **(5)** Storage/detention: modified Puls method requires table lookup for calculation of outflow. **(6)** Surface quality, quality routing, and treatment: algebraic equations, no iterations required once flows and conduit volumes are known.

Computational Status

(1) Coded in FORTRAN-77, approximately 25,000 statements long. **(2)** Has been run on IBM, UNIVAC, CDC, Amdahl, VAX, Prime, Harris, Boroughs, and other mainframe and minicomputers. **(3)** Personal computer version for IBM-PC compatibles uses FORTRAN code; requires 640 KB of memory, hard disk is desirable, executable and source code available. **(4)** May be run in modular form (RUNOFF, TRANSPORT, EXTRAN, and/or STORAGE/TREATMENT, plus executive and service routines (i.e., plotting, file combining, statistics, data input). **(5)** Largest block requires about 125,000 words or 500K bytes of storage. **(6)** Available on a magnetic tape or floppy disks. **(7)** Requires up to eight off-line storage files.

Input Data Requirements

(1) Historical or synthetic precipitation record; uses National Weather Service precipitation tapes for continuous simulation. **(2)** Monthly or daily evaporation rates. **(3)** For snowmelt: daily max-min (continuous) or time-step (single-event) temperatures, monthly wind speeds, melt coefficients and base melt temperatures, snow distribution fractions and areal depletion curves (continuous only), other melt parameters. **(4)** Surface quantity: area, imperviousness, slope, width, depression storage, and Manning's roughness for pervious and impervious areas; Horton or Green-Ampt infiltration parameters. **(5)** Subsurface quantity: porosity, field capacity, wilting point, hydraulic conductivity, initial water-table elevation, ET parameters; coefficients for ground-water outflow as function of stage-and-tail-water elevation. **(6)** Channel/pipe quantity: linkages, shape, slope, length, Manning's roughness; EXTRAN also requires invert and ground elevation, storage volumes at manholes and other structures; geometric and hydraulic parameters for weirs, pumps, orifices, storages, etc.; infiltration rate into conduits. **(7)** Storage/sedimentation quantity: stage-area-volume-outflow relationship, hydraulic characteristics of outflows. **(8)** Surface quality (Note: Several parameters are optional, depending upon methods used): land use; total curb length; catch-basin volume and initial pollutant concentrations; street weeping interval, efficiency, and availability factor; dry days prior to initial precipitation; dust/dirt and/or pollutant fraction parameters for each land-use or pollutant rating curve coefficient; initial pollutant surface loadings; exponential and power washoff coefficients; erosion parameters for Universal Soil Loss Equation, if simulated. **(9)** Dry-weather flow constant or on basis of diurnal and daily quantity/quality variations, population density, other demographic parameters. **(10)** Optional particle-size distribution, Shields parameter and decay coefficients for channel/pipe quality routing and

(continued)

Table 2-1. Continued

scour/deposition solution routine. **(11)** Storage/treatment: parameters defining pollutant removal equations; parameters for individual treatment options, e.g., particle-size distribution, maximum flow rates, size of unit, outflow characteristics, optional dry-weather flow data when using continuous simulation. **(12)** Storage/treatment costs: parameters for capital and operation and maintenance costs as function of flows, volumes, and operating time. **(13)** Data requirements for individual blocks much less than for run of whole model; large reduction in data requirements possible by aggregating (lumping) of subcatchments and channel/pipes, especially useful for continuous simulation. **(14)** Metric units optional for all input and output.

Ease of Application

(1) Nonproprietary model available from EPA, Athens, GA. **(2)** Updated user's manual and thorough documentation of most routines published as EPA reports; no one report covers all model aspects. **(3)** Test cases documented in several EPA and other reports. **(4)** Short course proceedings also useful for model applications. **(5)** U.S. and Canadian user groups with newsletters and semiannual meetings permit publication of changes. **(6)** Due to its age (originally published in 1971), availability, and documentation, examples of SWMM usage are widely available in the literature; bibliography available. **(7)** Frequent model update/corrections/improvements are often difficult to learn about; new model released approximately on a biannual basis. **(8)** Size of model most frequent deterrent to use; however, see item 13 under "Input Data Requirements" (preceding). **(9)** Initial model setup often moderately difficult due to size. **(10)** Model supported by EPA Center for Water Quality Modeling, Athens, GA.

Output and Output Format

(1) Input data summary including precipitation. **(2)** Hydrographs and pollutographs (concentrations and loads versus time) at any point in system on time-step or longer basis; no stages or velocities printed. **(3)** EXTRAN also outputs elevation of hydraulic grade line. **(4)** Surcharge volumes and required flow capacity; TRANSPORT Block will resize conduits to pass required flow (optional). **(5)** Stage, discharge, and soil moisture content for subsurface routing in RUNOFF Block. **(6)** Removal quantities in storage/treatment units, generated sludge quantities. **(7)** Summaries of volumes and pollutant loads for simulation period, continuity check, initial and final pounds of solids in conduit elements. **(8)** Daily (optional), monthly, annual, and total summaries for continuous simulation, plus ranking of 50 highest time-step precipitation, runoff, and pollutant values. **(9)** Line-printer plots of hyetographs, hydrographs, and pollutographs. **(10)** Costs of simulated storage/treatment options. **(11)** Statistical analysis of continuous (or single-event) output for event separation, frequency analysis, moments and identification of critical events.

Linkages to Other Models

(1) Linkage provided to EPA WASP and DYNHYD receiving-water–quality models. **(2)** Individual blocks and the total SWMM model have been linked to the HEC STORM model, the QUAL-II model, simplified receiving-water models, and others. **(3)** Individual blocks (e.g., RUNOFF Block) have been altered by various groups.

Personnel Requirements

(1) Civil/environmental engineer familiar with urban hydrological processes for data reduction and model analysis. **(2)** Computer programmer for model setup and establishment of off-line files on mainframes.

Costs

(1) Model and documentation available at little or no charge from EPA Center for Exposure Assessment Modeling, Athens, GA 30613; also available for nominal charge from Department of Environmental Engineering Sciences, University of Florida, Gainesville, FL 32611. **(2)** Data assembly and preparation may require multiple man-weeks for a large catchment or urban area. **(3)** Example mainframe computer execution costs given in user's manual, on the order of $20 for a RUNOFF and TRANSPORT run for a single-storm event with about 50 subcatchments and channel-pipes; use of EXTRAN can be more costly (greater than $100 per run) due to short time-step; continuous simulation (hourly time-step) of one subcatchment with snowmelt for two years costs about $20. **(4)** Microcomputer execution times depend on machine; execution times on AT-compatible approximately 50 times longer than for mainframe. **(5)** Extensive calibration may be required to duplicate measured quality results, quantity calibration relatively simple. **(6)** National Weather Service precipitation tapes for continuous simulation cost about $200 for at least a 25-yr hourly record for all stations in one state.

Model Accuracy

(1) Quantity simulation may be made quite accurate with relatively little calibration. **(2)** Quality simulation requires more extensive calibration using measured pollutant concentrations; quality results will almost certainly be very inaccurate without local measurements. **(3)** EXTRAN accurately simulates backwater, flow reversal, surcharging, pressure flow; TRANSPORT routines may be used at less cost if these conditions are not present. **(4)** Sensitivity to input parameters depends upon schematization; however, surface-quality predictions are most sensitive to pollutant loading rates.

---

surface runoff can occur. In a typical model application, this process is repeated for a number of subcatchments.

Mathematically, surface runoff is generated by two equations. One is a continuity equation that keeps track of the volume or depth of water on the surface (i.e., in the reservoir). The other is Manning's equation, which governs the rate

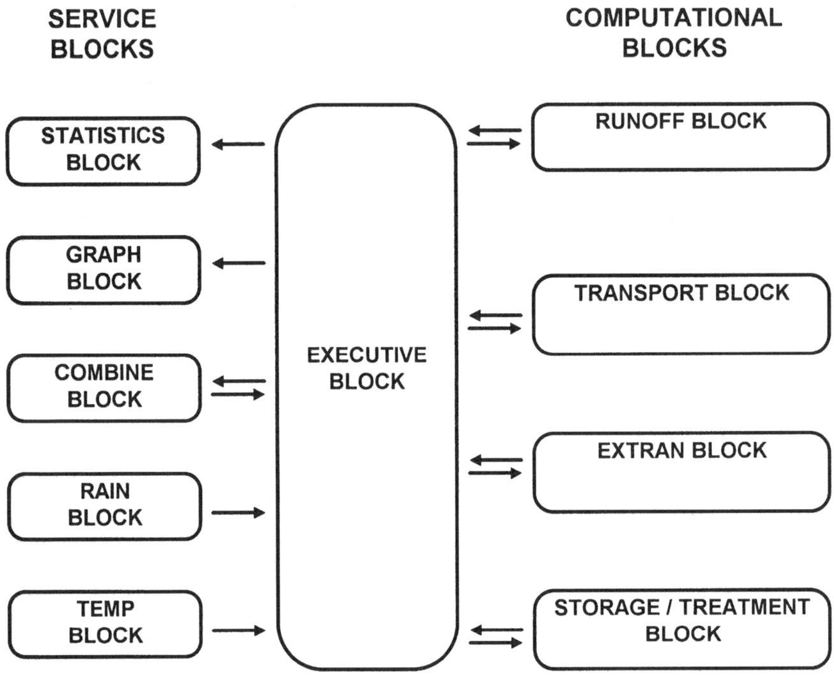

Figure 2-8. Operational schematic for Storm Water Management Model (Huber and Dickinson 1988).

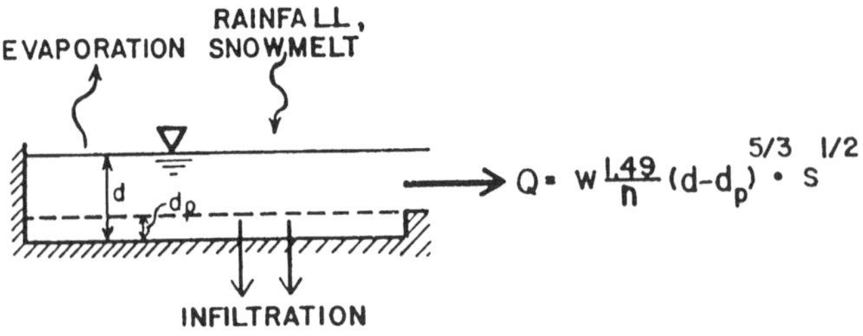

Figure 2-9. Nonlinear reservoir representation of a subcatchment—SWMM (Huber and Dickinson 1988).

of surface runoff (i.e., the discharge rate from the subcatchment) as a function of the depth of flow above the depression storage level. A complete description of this conceptualization of the surface runoff process can be found in the SWMM user's manual (Huber and Dickinson 1988).

The most confusing aspect of RUNOFF is the area over which surface runoff occurs—the "width of overland flow," as it is called. Again, using the reservoir analogy, this area is similar to the length of a weir or spillway. An idealized view is shown in Figure 2–10. In this schematic, surface runoff is discharged to a drainage channel located in the center of the subcatchment. If the two halves are symmetrical in all aspects, the total length of overland flow is twice the length of the channel. Of course, this idealized case never occurs, but it demonstrates the concept. This parameter, which must be provided by the user, is

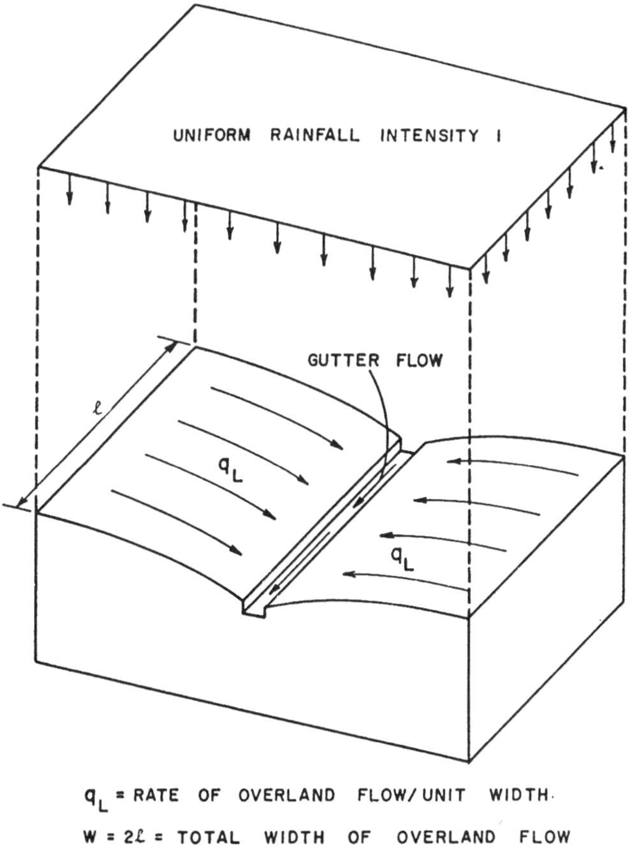

Figure 2–10. Idealized subcatchment illustrating the width of overland flow—SWMM (Huber and Dickinson 1988).

not readily estimated and is typically used as a calibration parameter. This is especially true when adjusting the shape of the runoff hydrograph (the wider the subcatchment, the faster the water can be discharged).

RUNOFF has the ability to route flows through gutters and pipes using the nonlinear reservoir technique. However, only three cross-sectional shapes are modeled, and the routing technique is not adequate for many applications. The sewer-routing algorithms in the TRANSPORT and EXTRAN Blocks are usually employed for this purpose. However, RUNOFF can be useful in modeling smaller lateral sewers, while TRANSPORT or EXTRAN focuses on the trunk sewers. (The routing characteristics of these three blocks are discussed in the next section.) The surface flows generated by RUNOFF are concentrated at nodes (i.e., the flows are not distributed along gutters as implied by Figure 2–10). The width of overland flow is used as a computational tool, but the flow is not distributed over this distance.

The accumulation of pollutants on the land surface is simulated in a number of ways. Pollutants can accumulate as dust and dirt on streets or as a simple areal load. The rate of accumulation may be linear or nonlinear. The accumulated load is washed off the land surface by a power function that relates the rate at which the load is washed off to the runoff rate and the amount remaining (and some other factors) or by a power function that relates the washoff rate to the runoff rate only. A substantial number of data are needed for the pollutant buildup and washoff routines, and the data-entry process is confusing. Whether the resulting accuracy is worth the effort (see earlier discussion) is still questionable.

There are many other capabilities not discussed here, most notably snowmelt simulation. The RUNOFF Block consumes a considerable portion of the SWMM user's manual, making it seem more profound and difficult than it is. Recall that the heart of the block is a very simple nonlinear reservoir representation of the surface runoff process.

TRANSPORT Block

The TRANSPORT Block routes flows and pollutant loads through a sewer system. These flows and loads may be generated by the RUNOFF Block or some other program and input to points throughout the system. TRANSPORT also has the ability to generate dry-weather or sanitary sewage flows for routing through a sewer system. The user may also introduce constant flows or time-varying hydrographs as inflows to various points in the system. The sewer system is viewed as a series of elements or links and nodes. Elements may be nodes (also referred to as junctions) such as manholes, pump stations, and overflow structures, or conduits linking the nodes. Conduits may have 15 different regular cross-sectional shapes (e.g., circular or rectangular) supplied by the program, an irregular or "natural" shape, or two supplied by the user. The different element types are listed in Table 2–2. The sewer system or network is specified by indicating each element with its associated upstream element(s). The elements are identified by user-supplied numbers.

Ideally, flow can be represented by two differential equations: the continuity and momentum equations, or, as they are sometimes known, the Saint-Venant equations. Unfortunately, these equations are difficult to manipulate and simpli-

Table 2–2. Element Types in Transport Block—SWMM (Huber and Dickinson 1988)

NTYPE	CONDUITS/CHANNELS
1	Circular
2	Rectangular
3	Phillips standard egg-shaped
4	Boston horseshoe
5	Gothic
6	Catenary
7	Louisville semielliptic
8	Basket-handle
9	Semicircular
10	Modified basket-handle
11	Rectangular, triangular bottom
12	Rectangular, round bottom
13	Trapezoid
14	Parabolic
15	Power function
16	HEC-2 format—natural channel
17, 18	User-supplied
	NONCONDUITS
19	Manhole
20	Lift station
21	Flow divider
22	Storage unit
23	Flow divider—weir
24	Flow divider
25	Backwater element

fications are often desirable. TRANSPORT uses a simplified version of the momentum equation in which flow is a simple function of depth. This kinematic approximation poses a severe restriction in that waves are propagated only in the downstream direction. Therefore, backwater effects, surcharging (the transition between open channel to pressure flow), looped connections, and pressure flows are either not simulated or are handled in very rudimentary ways. Probably the best way to understand the way the model operates is to view the system as a cascade of conduits with downstream conduits having no effect on upstream conduits. Simple representations of flow-diversion devices (e.g., overflow structures) are also provided. The original user's manual can be consulted for more details on the theoretical and mathematical details (Metcalf and Eddy, Inc., et al. 1971).

It is especially important to understand how the TRANSPORT Block behaves when it encounters surcharge conditions. Flows exceeding the full-flow conduit capacity of a conduit are stored at the upstream end of the conduit (at a node) and released when the capacity again becomes available. Hydrographs passing through such a conduit become "clipped" and, as a result, potential surcharge problems at downstream conduits may be masked. In short, the TRANSPORT Block is not an appropriate model for situations where appreciable surcharge problems are expected.

For comparison, the flow-routing capabilities of the RUNOFF, TRANSPORT, and EXTRAN Blocks are summarized in Table 2–3. As discussed earlier, RUNOFF has a very limited routing algorithm based on a linear reservoir representing a limited range of conduit types. TRANSPORT's abilities have just been outlined. EXTRAN has a more complete representation of the hydraulics of sewer systems, but it comes at a price (see later discussion).

The TRANSPORT Block has very simple routines to estimate infiltration into the sewer system. (This is, of course, different from ground-surface infiltration.) In fact, the user essentially supplies the infiltration amounts. The routine is not very useful, and the user can almost as easily enter infiltration flows at various nodes in the system. Unfortunately, there is no provision for generating sewer infiltration from the flows routed through the subsurface routing algorithms in RUNOFF, except through connection to the gutters/conduits modeled in RUNOFF.

TRANSPORT contains a fairly detailed routine for estimating dry-weather flows, which are estimated on the basis of land use, population, income levels, and a host of other factors. The amount of data required is substantial. It may be easier to obtain actual measurements of dry-weather flow or estimates from infiltration/inflow studies (or similar studies) and input these values directly into the system at various nodes. Pollutants are routed very simply through the system by treating each conduit as a completely mixed reactor with simple first-order decay. Solids scour and deposition are also simulated.

Table 2-3. Routing Characteristics of RUNOFF, TRANSPORT, and EXTRAN Blocks—SWMM (Huber and Dickinson 1988)

	Blocks		
	RUNOFF	TRANSPORT	EXTRAN
1. Flow routing method	Nonlinear reservoir, cascade of conduits	Kinematic wave, cascade of conduits	Complete equations, interactive conduit networks
2. Relative computational expense for identical network schematizations	Low	Moderate	High
3. Attenuation of hydrograph peaks	Yes	Yes	Yes
4. Time displacement of hydrograph peaks	Weak	Yes	Yes
5. In-conduit storage	Yes	Yes	Yes
6. Backwater or downstream control effects	No	No*	Yes
7. Flow reversal	No	No	Yes
8. Surcharge	Weak	Weak	Yes
9. Pressure flow	No	No	Yes
10. Branching tree network	Yes	Yes	Yes
11. Network with looped connections	No	No	No
12. Number of preprogrammed conduit shapes	3	16	8
13. Alternative hydraulic elements (e.g., pumps, weirs, regulators)	No	Yes	Yes
14. Dry-weather flow and infiltration generation (base flow)	No	Yes	Yes
15. Pollutograph routing	Yes	Yes	No
16. Solids scour/deposition	No	Yes	No
17. User input of hydrographs/pollutographs	No	Yes	Yes

*Backwater may be simulated as a horizontal water surface behind a storage element.

In summary, the TRANSPORT Block effectively routes flows through simple cascading sewer systems. It is inappropriate, however, for systems with complicated hydraulic connections and surcharging pipes.

EXTRAN Block

The EXTRAN Block extends some of the capabilities of the TRANSPORT Block (although it has its own developmental history). It does this by retaining the complete versions of the two equations governing flow. Thus, it is capable of propagating waves upstream and, in turn, simulating backwater effects, surcharging, pressure flow, looped connections, and so on. It has a history of being an unstable, difficult model that requires a fair amount of pampering (although the latest version may have addressed this tendency). The routing of pollutant loads is not simulated. A full description can be found in a document that accompanies the main SWMM user's manual (Roesner et al. 1988).

As in the TRANSPORT Block, the sewer system is viewed as a network of links and nodes (or, collectively, elements). The catalog of element types is not as extensive as that of the TRANSPORT Block (see Table 2–4), and the method of linking the system is slightly different. Instead of specifying only the elements upstream of each component, upstream and downstream nodes are assigned to each link (or conduit). Inflows to the system occur at the nodes and may be entered directly by the user or come from other programs, such as the RUNOFF Block, through an interface file. There are no routines for estimating sewer infiltration or dry-weather flows.

In order to keep stable the numerical techniques used to solve the governing equations, some attention must be paid to the size of the time-step. The user's manual gives clear instructions for estimating the proper values. Typically, the time-step is less than 30 seconds. By comparison, the TRANSPORT Block can usually operate satisfactorily with time-steps of many minutes. The short time-step results in relatively long run times.

The effort to produce satisfactory EXTRAN results can be considerable. EXTRAN is a powerful routine that requires a good deal of user skill. Nevertheless, systems that are complicated and/or subject to surcharging require its sophistication. Be on guard!

STORAGE/TREATMENT Block

The STORAGE/TREATMENT Block is designed to route flow and pollutant loads through a storage and/or treatment facility. The user is given a great deal of flexibility by the block's ability to connect as many as five units in a variety of networks. Each unit may be given detention (or storage) characteristics or be a

Table 2-4. Element Types in EXTRAN—SWMM (Roesner et al. 1988)

Element Class	Types
Conduits or Links	Rectangular Circular Horseshoe Egg-shaped Basket-handle Trapezoid Power function Natural channel (irregular cross section)
Junctions or Nodes (Manholes)	—
Diversion Structures	Orifices Transverse weirs Side-flow weirs
Pump Stations	On-line or off-line pump station
Storage Basins	On-line, enlarged pipes or tunnels On-line or off-line, arbitrary stage-area relationship
Outfall Structures	Transverse weir with tide gate Transverse weir without tide gate Side-flow weir with tide gate Side-flow weir without tide gate Outfall with tide gate Free outfall without tide gate

simple flow-through device. If a unit is modeled as a detention unit, flows are routed through the unit with the modified Puls method, which requires information regarding the relationships between stage, storage, and discharge. Pollutants are routed by assuming that the unit behaves as a completely mixed reactor or a plug-flow reactor. Pollutants are removed through the use of removal equations or discrete particle settling. In the former, the program provides several state variables (e.g., detention time, inflow rate, etc.) around which the user can build a wide range of removal equations. In the latter method, each pollutant is removed based on its association with particles of various settling velocities or sizes and specific gravities (e.g., 20% of the BOD load is associated with particles that have a given settling velocity). When a unit is defined as a simple flow-through device, flow is routed without delay. Pollutant removal is simulated with a removal equation (again, built by the user with a number of state variables provided by the program).

The STORAGE/TREATMENT Block uses relatively simple representations of pollutant routing and removal processes. It is not intended for the detailed simulation of steady-state unit operations like those found in sanitary sewage-treatment plants. There are many models that simulate these processes in great detail. This block is intended to deal with dynamic stormwater flows and pollutant loads and to give the user an idea of how a storage/treatment facility will respond under these conditions. As such, the pollutant removal modeling techniques are fairly simple in order to keep the model tractable. Those wanting to simulate the activated sludge processes in a sanitary sewage treatment plant, for example, will be sorely disappointed.

Receiving-Water Modeling

Earlier versions of SWMM included an algorithm known as the RECEIVE Block, which is capable of routing flows and pollutant loads through receiving waters such as rivers, streams, and estuaries (Metcalf and Eddy, Inc., et al. 1971; Huber et al. 1975). The latest version of SWMM does not contain the RECEIVE Block. However, there are provisions in two EPA receiving-water models, DYNHYD and WASP, for interfacing with SWMM (Ambrose et al. 1986).

EXECUTIVE Block

The EXECUTIVE Block acts as the "nerve center" of SWMM. It directs the order in which the other blocks are executed and provides an interface between them, as well as external programs and data files. The interface files (containing the information that must be transferred between blocks) have a standardized format to ease the interaction between blocks and external programs. The SWMM has been successfully interfaced with several other models.

GRAPH Block

The GRAPH Block provides a limited ability to plot actual and simulated hydrographs, pollutographs, and so on.

COMBINE Block

The computational blocks of SWMM have limitations on the number of elements (pipes, subcatchments, etc.) they can simulate. The COMBINE Block allows the modeling of larger areas by collating and combining interface files. Collation refers to the block's ability to add flows and pollutant loads generated by two other blocks (or two runs of the same block) at common nodes (i.e., manholes, inlets, etc.). To combine means that the output from two blocks (at two different nodes) is added and placed at one node.

RAIN Block

The RAIN Block processes lengthy precipitation records for use by the RUNOFF Block. These records are most often those provided on disk by the National Weather Service, but user prepared records may also be processed. RAIN also incorporates an algorithm to perform storm-event analyses. This routine is based on the well-known SYNOP program developed by Hydroscience, Inc. (1976, 1979).

TEMP Block

The TEMP Block processes daily temperature, evaporation, and wind speed data for use by the RUNOFF Block. These records are most often those provided on disk by the National Weather Service, but user-prepared records may also be processed.

STATISTICS Block

The STATISTICS Block performs a limited number of statistical operations on a sequence of events (runoff events, rainfall events, overflow events, etc.). A variety of quantity and quality parameters may be analyzed. Output can include (at the user's option) a table showing the sequential series of events; a table listing the magnitude, return period, and frequency of the events; graphs showing magnitude versus return period and frequency; and the first three moments of the event data. A user-defined minimum interevent time is used to separate events.

Case Studies

Overview

Combined Sewer Overflow (CSO) modeling techniques require both a statistical evaluation of the entire rainfall record as well as the development of CSO characteristics from specific "design" storms. The highly variable nature of rainfall and limitations in the long-term data available presents a unique challenge to the use of mathematical modeling for abatement evaluations. The evaluation of the major abatement alternatives such as storage and high-rate treatment requires the availability of long-term rainfall data and design storms, respectively.

The Environmental Protection Agencies (EPA) Storm Water Management Model (SWMM) has been used extensively to model the existing Syracuse CSO system in Onondaga County, New York, which is further described in chapter 5. The combined sewer tributary area is shown in Figure 2–11. The model was calibrated based on a series of flow and rainfall measurements throughout the collection system. Once calibrated, the model was then used to project design storm conditions for each overflow location. The design storms were used to project abatement scenarios for high-rate treatment alternatives. The use of continuous simulation was also employed to project volume requirements on a recurrence interval basis for storage alternatives.

Model Development

Version 4 of the SWMM model was used for the results of the most current study as discussed in this book. Due to the size of the drainage areas and resulting collection system, two separate models were developed for this study. The RUNOFF Block was used to generate surface runoff for each of the combined sewered areas tributary to the overflows.

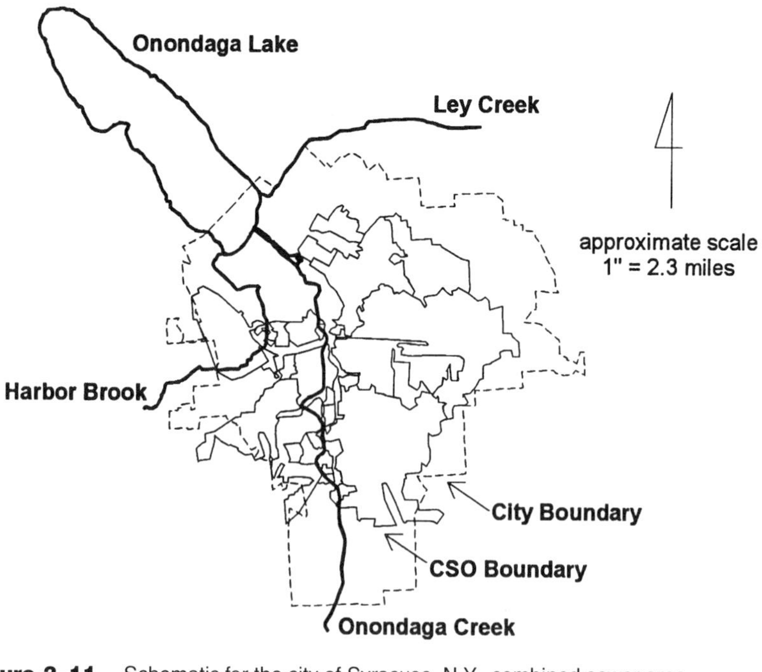

Figure 2–11. Schematic for the city of Syracuse, N.Y., combined sewer area.

The EXTRAN Block was used to model the sewer network. EXTRAN was used due to the extensive surcharging that exists during wet-weather events. Models were calibrated based on several storms that were monitored throughout the basin. The model was constructed to include the major interceptor and major trunk sewers, regulator structures, and one pipe segment upstream of each regulator structure. Two separate models were developed for each of two major drainage systems, namely, the Harbor Brook Intercepting sewer system and the Main Intercepting sewer system. The models included approximately 450 pipe segments. At the time of the development, the model was not capable of handling more than 400 pipe segments, and it was necessary to model the two systems separately. Although the two interceptors are both tributary to the Metropolitan Sewerage Treatment Plant, they were found not to be influenced by each other hydraulically. Thus, it was determined that separate models could be developed and run independently.

Average daily flow was distributed throughout the model network based on area weighting of the average daily flow. Diurnal changes in flow were not simulated since the model results are not sensitive to such small changes in flow as compared to the very large flows that result from storm conditions. The time-step selected for the EXTRAN block was 15 seconds. This value represented a good compromise between stability considerations and run times.

Model Calibration

The RUNOFF model was calibrated based on a total of nine monitored locations (CSOs). The sites represented approximately 2,307 acres of tributary area or 35% of the total combined sewer system. Although several storms were monitored, only two rainfall events were used for calibration purposes due to their peak intensity, total volume, and duration. One of the sites used for calibration is depicted in Figure 2–12. This figure illustrates the results of calibration for CSO 066 for two storms that occurred on 9/8/87 and 9/12/87. The tributary area to this overflow included 114 acres at approximately 70 percent imperviousness. The storm of 9/12/87 exhibited a series of high intensity rainfall periods that varied significantly across the entire CSO tributary area. These types of storms are sometimes difficult to calibrate due to their uneven distribution; however, these higher intensity storms have a beneficial aspect to them in that they include intensity characteristics more representative of design conditions.

Model Projections

Once calibrated, the models were then used to project flow conditions for various design storms. These flow projections were coupled with average-pollutant concentration data sampled throughout the collection system to project pollutant

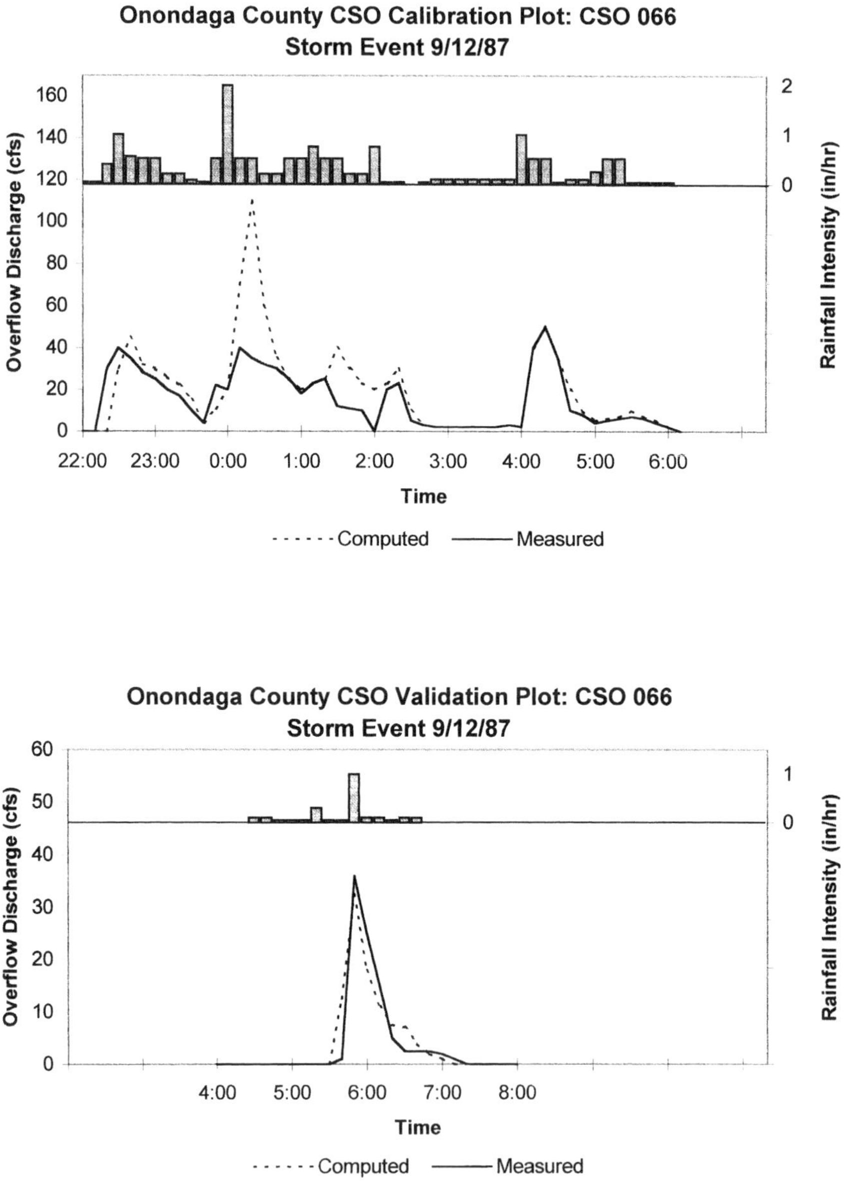

Figure 2-12. Calibration results of September 12, 1987, storm: Onondaga County, N.Y.

loadings for a variety of conditions and abatement alternatives. The final abatement alternative included regional high-rate treatment at four locations throughout the city.

The design storms that were used were developed based on the Keifer-Chu approach (also known as the Chicago Method). The one-year, two-hour design storm for Syracuse is shown in Figure 2–13. In addition, peak flow rates were identified for the two-, five-, and ten-year storms. Figure 2–14 depicts the results of the one-year design storm for the Midland Avenue regional treatment facility.

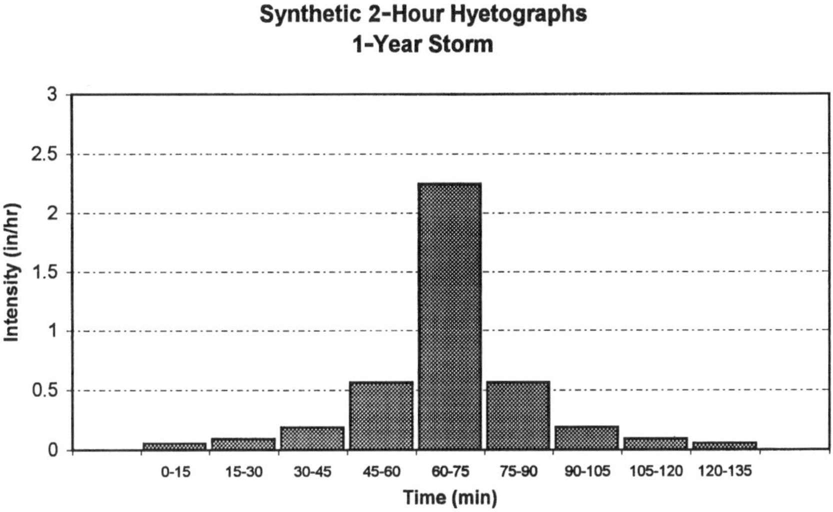

Figure 2–13. Design-storm hyetograph of Onondaga County, N.Y.

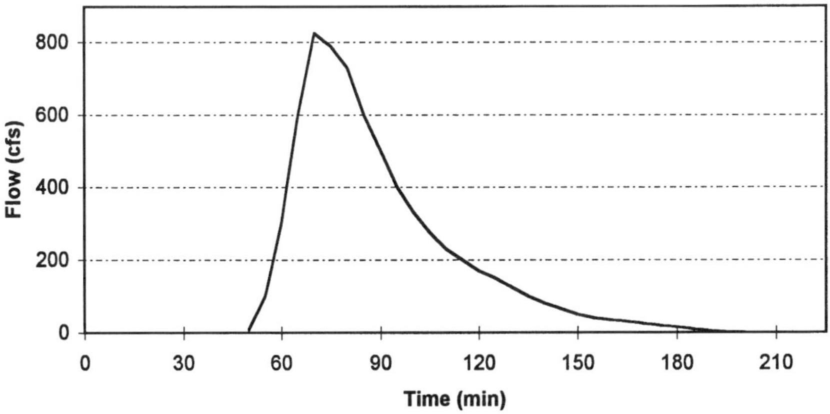

Figure 2–14. One-year storm influent hydrograph: Midland CSO, Onondaga County, N.Y.

Flows from several overflow locations in the proximity of the proposed facility were consolidated into this one treatment facility.

Continuous Simulation

The use of continuous simulation was also employed to identify the volumes associated with the annual discharge of CSO throughout the system. In addition, volumes were identified for the one-year return storm.

The TRANSPORT Block was utilized for the purposes of developing a sewer network for use in continuous simulation. The TRANSPORT Block allows the model to simulate several years of rainfall data to identify annual CSO characteristics. The EXTRAN block, which was used to identify storm-specific conditions, is not designed to run several years of rainfall data due to the computational complexity of the model.

The RUNOFF model was used to generate runoff hydrographs as input for the TRANSPORT models. Thirty years of hourly precipitation data were simulated. The resulting hydrograph was then analyzed for annual discharge volumes and the volumes for specific recurrence intervals (i.e., once per year).

Based on the long-term analysis, the one-year recurrence volume was calculated as 118 million gallons (MG) for the Onondaga Creek Basin. In comparison, the one-year design storm volume resulted in a discharge of 48 MG. It is also interesting to compare the one-year recurrence peak flow rates for the two referenced approaches. The total system, one-year peak rates for the Onondaga Creek Basin represent 1400 and 2000 million gallons per day (MGD) for the continuous simulation and design storm approaches, respectively.

Conclusion

The results of the preceding analyses allowed for a more precise comparison of abatement alternatives. The summary of volumes and peak rates calculated for the Onondaga Creek Basin in Syracuse using the different modeling approaches is shown in Table 2–5. The table illustrates the importance of us-

Table 2–5. Comparison of Results of Different Modeling Approaches

Modeling Method	One-Year Recurrence Volume (MG)	One-Year Recurrence Peak Rate (MGD)
Continuous Simulation	118	1,400
Design Storm	48	2,000

ing the appropriate modeling method depending upon the evaluation objective. The use of continuous-simulation modeling results in a volume of 118 MG for a one-year storm; any storage evaluations should be based on this value. If design-storm modeling had been used, storage requirements would have been underestimated by more than 50%. Conversely, continuous-simulation modeling would have yielded a one-hour peak rate that is 30% less than the 15-minute peak rate from the design-storm modeling. Thus a high-rate facility would have been undersized if continuous simulation had been utilized.

Concluding Remarks

The purpose of this chapter is to present a brief, clear discussion of the use of mathematical models to simulate the behavior of combined sewer systems. Models are very helpful in understanding, projecting, and abating CSO problems, but it should be clear that modeling must be part of an integrated attack on a problem, not its focus. A model should substantiate more than shape the expert's judgement. Much of the chapter focuses on the use of the EPA Storm Water Management Model. This was not intended to minimize the potential of other models. SWMM is a good, well-documented model, but it is not the only model capable of the task.

Fortunately, information about combined sewer modeling is plentiful, and new information appears regularly. Some of the more significant sources are listed below:

- Reviews of SWMM-related literature prepared by Huber et al. 1995 and James 1993. Both of these reviews are excellent.
- Annual literature review published by Water Environment Research (formerly the *Journal of the Water Pollution Control Federation*). The section on "Urban Runoff and Combined Sewer Overflows" is particularly useful.
- EPA Storm Water and Water Quality Model Users Group. This group meets twice per year, and its proceedings are published and readily available. Information on the Users Group can be obtained from the EPA Center for Exposure Assessment Modeling (CEAM), whose address was given earlier in this chapter.

Of course, there are a number of other sources of good information. Many appear on the Internet, and a search easily reveals them.

References: Chapter 2

Ambrose, R. B., S. B. Vandergrift, and T. A. Wool. September 1986. "WASP3, A Hydrodynamic and Water Quality Model-Model Theory, User's Manual and Programmer's Guide," EPA-600/3-86-034, USEPA, Athens, GA.

American Public Works Association. 1981. "Urban Stormwater Management," Special Report No. 49, APWA, Chicago, IL.

Bicknell, B. R., J. C. Imhoff, J. L. Kittle, A. S. Donigian, and R. C. Johanson. May 1993. "Hydrological Simulation Program—FORTRAN: User's Manual for Release 10," EPA-Draft, USEPA, Athens, GA.

Brandstetter, A. B. August 1977. "Assessment of Mathematical Models for Storm and Combined Sewer Management," EPA-600/2-76-175a, USEPA, Cincinnati, OH.

Chu, C. S., and C. E. Bowers. November 1977. "Computer Programs in Water Resources," WRRC Bulletin 97, Water Resources Research Center, University of Minnesota, Minneapolis, MN.

DiToro, D. M., and M. J. Small. "Stormwater Interception and Storage," *Journal of the Environmental Engineering Division,* ASCE, vol. 105, no. EE1 (February 1979): 43–54.

Donigian, A. S., Jr., and N. H. Crawford. February 1976. "Modeling Pesticides and Nutrients on Agricultural Lands," EPA-600/2-76-043, USEPA, Athens, GA.

Environmental Protection Agency. July 1983. "EPA Environmental Data Base and Model Directory," Information Clearinghouse (PM 211A), USEPA, Washington, D.C.

Geiger, W. F., and H. R. Dorsch. March 1980. "Quantity-Quality Simulation (QQS): A Detailed Continuous Planning Model for Urban Runoff Control: Vol. I, Model Description, Testing and Applications," EPA-600/2-80-011, USEPA, Cincinnati, OH.

Hall, M. J. 1984. "Urban Hydrology," Elsevier Applied Science Publications, New York.

Heaney, J. P., W. C. Huber, and S. J. Nix. October 1976. "Storm Water Management Model: Level I-Preliminary Screening Procedures," EPA-600/2-76-275, USEPA, Cincinnati, OH.

Heaney, J. P., and S. J. Nix. 1977. "Storm Water Management Model: Level I-Comparative Evaluation of Storage-Treatment and Other Management Practices," EPA-600/2-77-083, USEPA, Cincinnati, OH.

Howard, C. D. D. "Theory of Storage and Treatment-Plant Overflows," *Journal of the Environmental Engineering Division,* ASCE, vol. 102, no. EE4 (August 1976): 709–22.

Huber, W. C. 1985. "Deterministic Modeling of Urban Runoff Quality," in H. Torno, J. Marsalok, and M. Desbordes, eds., "Urban Runoff Pollution," NATO ASI Series, Series G: Ecological Sciences, vol. 10. New York: Springer-Verlag, 166–242.

———. June 1986. "Modeling Urban Runoff Quality: State-of-the-Art," B. Urbonas and L. A. Roesner, eds., "Proceedings of Engineering Foundation Conference on Urban Runoff Quality-Impact and Quality Enhancement Technology." New York: ASCE, 34–48.

Huber, W. C., and J. P. Heaney. 1980. "Operational Models for Stormwater Quality Management," M. R. Overcash and J. M. Davidson, eds., "Environmental Impact of Nonpoint Source Pollution." Ann Arbor, MI: Ann Arbor Science Publications. 397–444.

———. 1982. "Analyzing Residuals Discharge and Generation from Urban and Non-Urban Land Surfaces," in D. J. Basta and B. T. Bower, eds., "Analyzing Natural Systems, Analysis for Regional Residuals-Environmental Quality Management," Resources for the Future, Washington, D.C. Baltimore, MD: The Johns Hopkins University Press, 121–243. (This is also available as EPA-600/3-83-046, USEPA.)

Huber, W. C., and R. E. Dickinson. August 1988. "Storm Water Management Model, Version 4: User's Manual," PA-600/3-88-001a, USEPA, Athens, GA.

Huber, W. C., J. P. Heaney, M. A. Medina, W. A. Peitz, H. Sheikh, and G. F. Smith. March 1975. "Storm Water Management Model User's Manual—Version II," EPA-670/2-75-017, USEPA, Cincinnati, OH.

Huber, W. C., J. P. Heaney, S. J. Nix, R. E. Dickinson, and D. J. Polmann. 1984. "Storm Water Management Model User's Manual, Version IH," EPA600/2-84-109a, USEPA, Cincinnati, OH.

Huber, W. C., J. P. Heaney, and B. A. Cunningham. September 1985. "Storm Water Management Model (SWMM) Bibliography," EPA-600/3-85-077, USEPA, Athens, GA.

Hydrologic Engineering Center. August 1977. "Storage, Treatment, Overflow, Runoff Model, STORM," Generalized Computer Program 723-S8-17520, BEC, Corps of Engineers, Davis, CA.

Hydroscience, Inc. July 1976. "Areawide Assessment Procedures Manual," vols. I, II, and III, EPA-600/9-76-014, USEPA, Cincinnati, OH.

———. May 1979. "A Statistical Method for the Assessment of Urban Stormwater," EPA-440/3-79-023, USEPA, Washington, D.C.

James, W. "Introduction to the SWMM Environment," in W. James, ed., *New Techniques for Modeling the Management of Stormwater Quality Impacts.* Boca Raton, FL: Lewis Publishers, CRS Press, 1993.

Li, J. Y., and B. J. Adams. "Statistical Water Quality Modeling for Urban Runoff Control Planning," Water Science and Technology, vol. 29, no. 1–2 (1994): 181–90.

Metcalf and Eddy, Inc.; University of Florida; and Water Resources Engineers, Inc. 1971. "Storm Water Management Model," vols. I, II, III, and IV; EPA Reports 11024 DOC 07/71, 08/71, 09/71, 10/71; USEPA, Washington, D.C.

Mueller, J. A., and D. M. DiToro. 1983. "Combined Sewer Overflow Characteristics from Treatment Plant Data," EPA-600/S2-83-049, USEPA, Cincinnati, OH.

Roesner, L. A., R. P. Shubinski, and J. A. Aldrich. 1984. "Storm Water Management Model User's Manual, Version III: Addendum 1, EXTRAN," EPA-600/2-84-109b, USEPA, Cincinnati, OH.

Roesner, L. A., J. A. Aldrich, and R. E. Dickinson. 1988. "Storm Water Management Model User's Manual, Version IV: Addendum I, EXTRAN," EPA-600/3-88-001, USEPA, Athens GA.

Sonnen, M. B. "Urban Runoff Quality: Information Needs," *Journal of the Technical Councils,* ASCE, vol. 106, no. TC1 (August 1980): 29–40.

Terstriep, M. L., and J. B. Stall. 1974. "The Illinois Urban Drainage Area Simulator, 78 9 Control and Treatment of Combined Sewer Overflows ILUDAS," Bulletin No. 58, Illinois State Water Survey, Urbana, IL.

Whipple, W. Jr., N. S. Grigg, T. Grizzard, C. W. Randall, R. P. Shubinski, and L. S. Tucker. 1983. "Stormwater Management in Urbanizing Areas," Prentice-Hall, Englewood Cliffs, NJ.

Chapter 3

Receiving-Water Impacts

- John K. Marr, P.E.
  Associate Vice-President
  Limno-Tech, Inc.
  Ann Arbor, Michigan

- Paul L. Freedman, P.E.
  President
  Limno-Tech, Inc.
  Ann Arbor, Michigan

General Concerns

The previous chapter outlined methods to quantify combined sewer overflow (CSO) pollutants. This chapter discusses the impact these pollutants can have on water quality. The primary impetus for the control of CSOs is to protect the quality and uses of receiving waters. Therefore, any CSO control program and each different alternative must be evaluated in terms of how it reduces receiving-water problems and restores or maintains a beneficial use. To this end, general control policy and planning approaches are outlined, followed by the use of monitoring and modeling as means for assessment and the use of water quality benefits as a measure of performance for CSO controls. Assessment procedures are provided for streams, estuaries, and lakes. A discussion of the selection of design conditions is presented, and case examples are provided at the chapter's end.

National CSO Control Policy

In April 1994, the U.S. Environmental Protection Agency (U.S. EPA) published a Final CSO Control Policy in the *Federal Register.* The CSO policy is a national policy intended to engage regulators and CSO communities in coordinated planning to achieve cost-effective CSO controls. Ultimately, Long Term Control Plans (LTCPs) for all CSO communities as described in chapter 1 must meet water-quality standards and allow the appropriate beneficial uses of the CSO receiving waters. The policy is intended to consider site-specific factors and provide the flexibility to tailor CSO controls to local situations through the two approaches previously identified, namely, the presumption and the demonstration approach, which are restated for the purpose of this chapter.

- **Presumption Approach** This approach allows CSO agencies to presume that a specified end-of-pipe level of control will meet water-quality standards until or unless receiving-water monitoring proves otherwise. The end-of-pipe CSO control levels for the presumption approach are no more than an average of 4 to 6 overflows per year, treatment of 85% (by volume) of the collected CS flow (primary treatment plus some form of disinfection), or elimination by other means of the equivalent mass of CSO pollutant loading provided by the 85% volume control level. The presumption approach further requires water quality monitoring to confirm that the presumed effective CSO controls, when implemented, do indeed meet the applicable water quality standards.
- **Demonstration Approach** Intended for communities that either believe treatment less than the presumption level will meet water-quality standards or for those that cannot safely presume the attainability of water-quality standards under the presumption approach. With this approach, the LTCP must credibly demonstrate through monitoring and modeling that the proposed CSO controls, when implemented, will meet water-quality standards in the receiving water. Confirmatory monitoring after control implementation is also required.

With either LTCP approach, presumption or demonstration, it *must* be proven through follow-up monitoring that applicable water-quality standards are met in the receiving waters. Therefore, both approaches warrant characterization and forecasting of CSO impacts on water quality as described in this chapter.

The specific receiving-water impacts from CSOs vary, depending on the nature of the development in the sewered area, the size and frequency of overflows, the size and nature of the receiving stream, and the opportunity for uses of the receiving water. Depending on the particular combination of these factors, CSO impacts can be visible and intense. For example, the nature of the sewered area can determine the presence of toxic contaminants such as dissolved trace metals in the overflows, while the size and type of the receiving-water body will determine its ability to dilute and assimilate intermittent CSO discharges without toxicity. Therefore, receiving-water impacts must be evaluated based on each site's particular characteristics. In each study, CSO impacts and benefits can be evaluated with respect to three basic considerations:

- Water-quality changes and effects on aquatic organisms
- Public health risks
- Aesthetic deterioration

Each consideration affects the value of the receiving water for uses such as recreation, water supply, ecological habitat, and near-shore development. The significance of CSOs to the three major groups of considerations will vary, depending on the nature of the overflows and the receiving water. Figure 3–1 provides a qualitative assessment of the likelihood of significant CSO impacts on various types of receiving waters. Discussions of their general impacts are also provided in this chapter.

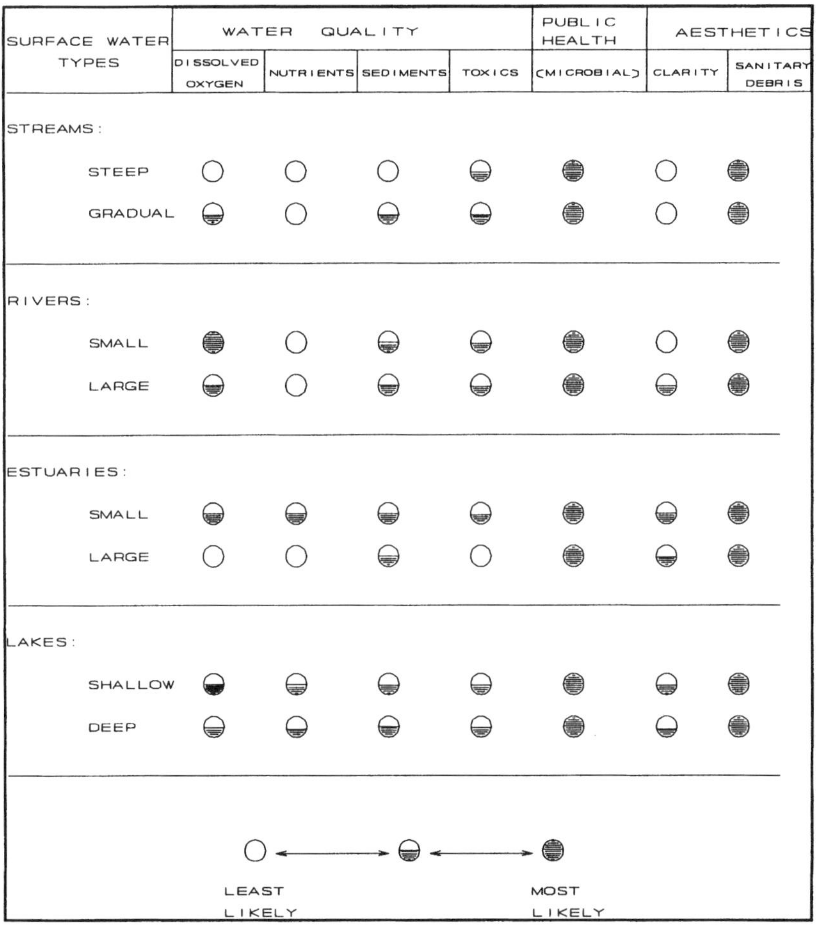

Figure 3–1. Qualitative assessment of the likelihood of significant CSO impacts on various receiving-water systems.

Water-Quality Changes and Effects on Aquatic Organisms

The CSO impacts on receiving waters depend on the rate and volume of CSO, the degree of mixing and dilution, and the receiving-water's assimilative capacity. Obviously, the larger the water body and the smaller the discharge, the better the likelihood of minimal impacts. For example, in some large rivers and estuaries, CSOs have minimal impacts because of rapid dilution and mixing. In contrast, small streams that receive CSO with little dilution can be severely impacted even for small storms. Therefore, each CSO study area requires individual monitoring and evaluation.

Most water-quality impacts are measured by a determination of the increases (or decreases) in receiving-water concentrations as they relate to a violation of water-quality criteria or standards. Such criteria are generally used to assess the potential for effects on aquatic organisms and impaired use. Although CSOs are intermittent, the resultant problems may not be temporary and can persist to varying degrees. Figure 3–2 provides preliminary guidance for the time frames to be considered for different water-quality problems. These different time frames can require various strategies for monitoring and calculating CSO impacts. Following are discussions for dissolved oxygen, nutrient enrichment, sediments, and toxicity problems.

Temporary depression of dissolved oxygen levels is one of the most common water-quality problems attributable to CSO. Short-term reductions due to BOD deoxygenation and ammonia nitrification can last from hours to days. Furthermore, oxygen reductions due to reduced sunlight and photosynthesis during a storm can compound the problem (and often confuse analyses). These dissolved oxygen reductions may impair aquatic growth or reproduction and, in severe situations, cause biologic mortality. The severity, frequency, and duration of CSO-caused dissolved oxygen reductions are key factors that cannot be generalized. Therefore, monitoring and modeling targeted at a three-to-six-day time frame following a storm can typically be used to quantify dissolved oxygen reductions. The long-term dissolved oxygen trend in CSO receiving waters has also been identified as one measure of CSO control performance (AMSA, January 1996).

Scientific criteria to define the biologic impacts of short-term reductions in dissolved oxygen are limited and generally inconclusive. Data show varying impacts for various species at a range of exposures. For example, large depressions (down to 1–2 mg/l) even very short in nature can cause mortality. In contrast, small reductions even for long periods have a small uncertain impact on growth and reproduction, especially when integrated with otherwise nonimpacted conditions. Figure 3–3 graphically summarizes a large body of scientific data on mostly long-term impacts. This figure can be used as guidance for preliminary assessment of the risk of CSO effects for warm-water systems. Existing state regulations and standards should always be consulted for their applicability to wet weather conditions and CSO control programs. The national CSO policy

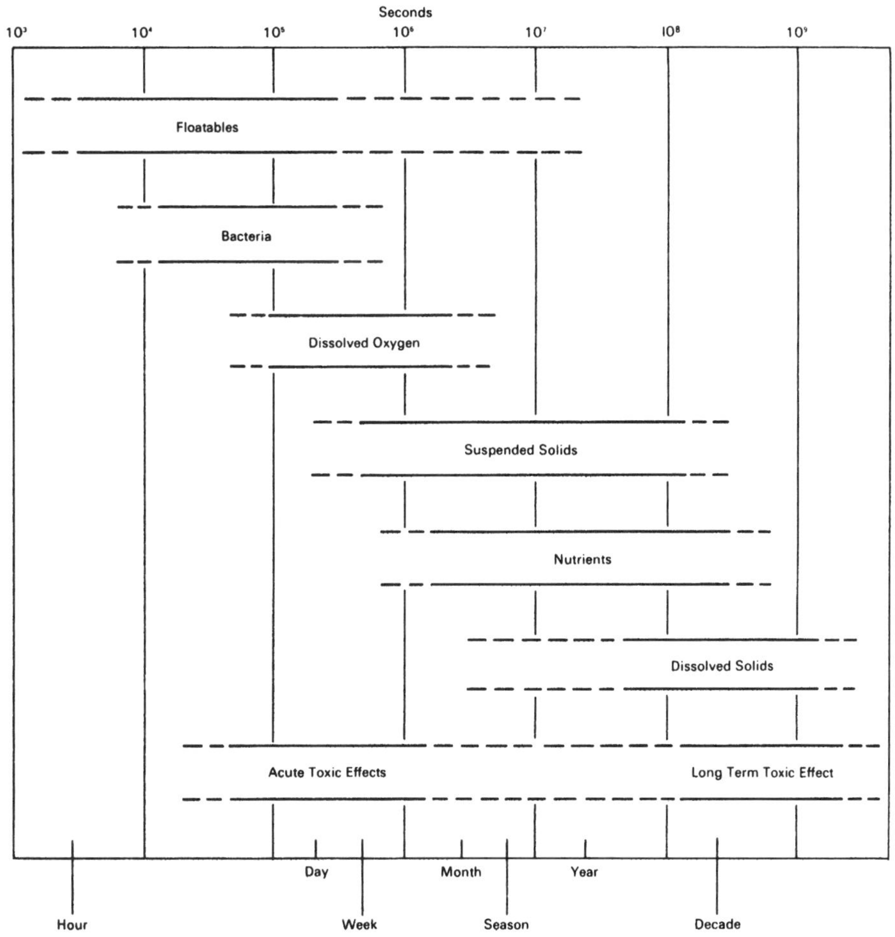

Figure 3–2. Time scales for storm-runoff and CSO water-quality problems.

provides flexibility to adjust the application of dissolved oxygen standards to reflect wet weather conditions (*Federal Register,* April 1996). This may be considered in situations where marginal and short-lived dissolved oxygen reductions caused by CSO occur below the standards, but where no detrimental effects are observed on aquatic organisms, including fish.

Another potential CSO problem is nutrient enrichment from phosphorus and nitrogen. Nutrient additions cause increased algae or aquatic weed growth. In turn, this may cause dissolved oxygen problems, reduce biologic diversity, worsen aesthetics, or impair use for water supply. The CSO contribution to this problem often cannot easily be distinguished from other sources of pollution.

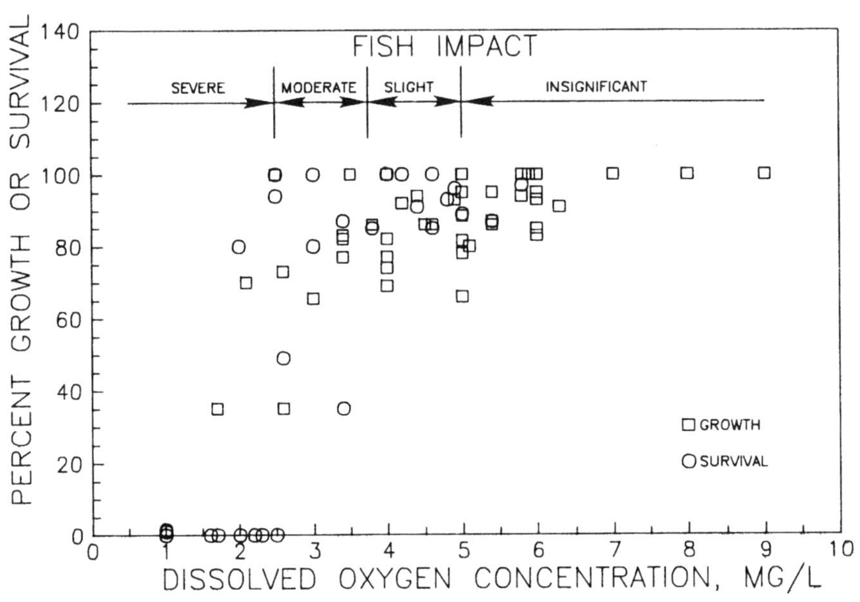

Figure 3–3. Effect of long-term, dissolved-oxygen concentrations on warm water fish growth and survival, James River, VA. (Limno-Tech, Inc., 1987).

Because algae and weed growth are seasonal, they tend to integrate the effects of all nutrient sources, especially those contributing during the peak growing periods. Hence, short-term monitoring and modeling may be ineffective in assessing CSO contributions to nutrient enrichment problems. In some systems, the long-term CSO contribution of nutrients is small compared to other sources such as wastewater and upstream nonpoint loadings, especially over the peak growing season. Assessments of the CSO importance to nutrient enrichment and eutrophication in slow-moving streams, estuaries, and lakes can be made by performing relative comparisons of seasonal nutrient loadings. In faster-moving streams, CSO nutrient contributions are typically flushed out before any impact can be realized. If quantitative evaluations of eutrophication are necessary, then a comprehensive program of seasonal monitoring and mathematical modeling is required.

In evaluating the impact of nutrient enrichment, objective criteria are often lacking. Algal and nutrient criteria are not common to all states and are often widely different. For example, state chlorophyll-a guidelines can range from 10 to 100 mg/l (API 1980). As a result, evaluations of CSO impacts on nutrient enrichment require site-specific consideration of existing and historical conditions, water-body uses, relevant standards and criteria, and the nature of the water body. In most cases, a seasonal or annual nutrient load balance is the most practical approach for assessment.

Another common problem from CSO is accumulated bottom-sediment contamination. The CSOs typically have high levels of suspended materials that settle to the bottom in receiving waters. The accumulation and persistence of CSO solids in sediments are typically long-term, not transient, problems. The results can be high sediment oxygen demand (SOD) due to biologic oxidation and accumulation of toxic substances in the sediments—primarily trace metals. These conditions are usually present in slower-moving rivers, estuaries, or lakes, but not in fast-moving streams where sediments are routinely scoured from the bottom. Sediment criteria and standards for evaluating the severity of a problem are uncommon. The EPA provides general guidelines for disposal of dredged sediments (EPA 1977), but beyond this, the significance of sediment contamination must usually be evaluated through consideration of effects on overlying waters, if any.

Initial efforts by the EPA toward formulation of sediment criteria and modeling methods are completed (EPA September 1993). However, this research is incomplete and not yet available for practical applications in CSO control evaluations. Sediment evaluations of CSO impacts generally require extensive monitoring, mathematical modeling, and comparisons with corresponding water-quality criteria. Similar to nutrients, bottom sediments arise from many contributing sources. Initial assessments of CSO contributions can often be derived by comparing the annual or seasonal inputs of relevant settleable materials from all sources. Though difficult to predict, improvements in sediment conditions and benthic communities can be used to assess the performance of CSO control programs (AMSA January 1996), as discussed in the performance measures section of this chapter.

The fourth general impact from CSO is receiving-water toxicity. In general, toxicity problems fall into two categories: (1) chronic or long-term exposure to toxicants causing reduced growth and reproduction, or (2) acute toxicity at higher concentrations causing short-term increases in mortality. Acute effects can often be seen as immediate fish kills or severely reduced biologic diversity. Chronic effects are more subtle and harder to identify but at times can be observed by lower productivity and biomass (numbers of organisms), by bioaccumulation of chemicals, or by reduced biological diversity. Fortunately, chronic effects are not often found because CSOs are usually intermittent and infrequent.

Toxic discharges from CSO can generally be considered a preliminary concern in industrialized areas that are combined sewered or if monitoring data for the sanitary sewage indicate potential toxicity. Pretreatment programs are usually a good source of information. The more common toxicity considerations are for ammonia and dissolved trace metals, but even these do not warrant consideration in many CSO impact evaluations because end-of-pipe concentrations are too low for concern. Preliminary screening for toxic effects can be performed using average CSO concentrations and peak hourly flow rates for calculating an instream impact using a range of expected dilution factors. If such considerations are relevant to a site, then existing state and EPA water-quality criteria are available for making preliminary assessments (EPA 1986).

Extreme care should be taken to consider the appropriate time frames and applicability of the criteria to wet-weather situations. In most cases, trace metals should be evaluated on the basis of the dissolved, bioavailable fractions in the CSOs and receiving waters. The Water Environment Research Foundation (Milne et al. 1995) is conducting ongoing research to evaluate the proper application of toxicity testing for intermittent CSO and storm water conditions. Preliminary results from this research indicate that wet-weather toxicity testing must consider the timing and exposure of organisms to CSO and storm-water pollutant emissions to be valid.

Public Health Risks

By design, CSOs include untreated sanitary wastes and contaminated storm runoff. Although human health toxicity is a remote concern, the primary public health issue from CSOs is exposure to risk of disease. Sanitary wastes carry human excrement that can include pathogenic bacteria and viruses. Human exposure to CSO-contaminated receiving-waters through wading, bathing, shellfish ingestion, and fishing therefore poses a risk of exposure to these pathogens and is a concern in nearly all types of receiving waters. The high levels of bacteria observed in CSOs can cause potential problem conditions in receiving waters even with very high dilution. The potential risk is typically greatest immediately after the overflow during the first flush of sanitary wastes and can diminish quickly due to bacterial die-off, settling, and dilution. Significant public health risks rarely persist more than two or three days after a storm overflow event. Overflows laden with bacteria are special concerns during warm weather when water contact recreation peaks and human exposure is greatest. However, bacterial death rates are high during warm weather, tending to reduce total exposure times.

Human health risk from CSOs is not well defined because it is difficult to measure all potential pathogens and even harder to define a link between exposure and human health. Most CSO studies use the presence of fecal coliform bacteria (or other enteric bacteria such as *Escherichia coli*) as indicators of the presence of sanitary waste. This, in turn, suggests the possibility for exposure to pathogenic bacteria or viruses. State guidelines and standards for contact recreation and water supply are common for fecal coliform and other related bacteria.

Generally, receiving-water concentrations above 200–1000 organisms/100 ml are considered to represent increased exposure to disease risk. Concentrations of indicator bacteria in CSOs usually number in the millions, and hence, dilution is rarely sufficient to eliminate all potential risk. The CSO risks to public health can be quantified by the frequency, severity, and duration of elevated indicator bacteria concentrations. Generally speaking, increased public health risks from CSOs are a very prevalent and potentially serious issue. At times, the concern may be only transient and local (by the outfall), but in many systems the concern may extend for miles and days beyond the discharge. **In all cases, CSO-caused risks to public health must be given the highest consideration.**

Inasmuch as bacterial indicators play a vital role in determining the CSO abatement needs of many communities, it is appropriate to review the history of the bacterial standards and the future directions we may be taking. Of all the water-quality parameters, bacteria comprise the only one in which an indicator serves as the actual criterion. This "guilt by association" plays perhaps the single most critical role in determining acceptability of the federal "swimmability" goal. As such, it should represent a well-defined, highly credible measurement. However, because of the concern for analytically handling pathogens directly, such is not the case. A brief history of indicator organisms is in order.

The first, most widely used indicator for fecal materials was the coliform bacterium. As early as 1934, a classification scheme was developed to categorize the quality of waters (W. J. Scott 1934). Though the studies were for Connecticut waters, the method was adopted by the American Public Health Association in 1943. Subsequently, one of the later studies conducted on the Ohio River by the U.S. Public Health Service found that an excess of gastrointestinal illness in swimmers occurred when the median total coliform density was 2,300/100 ml and greater (A. H. Stevenson 1953). In 1968, the U.S. Department of the Interior's National Technical Advisory Committee (NTAC) proposed the use of the total coliform (TC) indicator in determining swimming area suitability (U.S. Dept. of the Interior 1968). Shortly thereafter, the NTAC proposed the use of the fecal coliform (FC) bacterial indicator in lieu of the TC indicator because: (1) it was more indicative of fecal contamination, and (2) it was less subject to variation. (Even at that time, the confounding influence of storm runoff on receiving waters was evident.) The identification of the FC in the Ohio River study resulted in approximately 18% of the TC count, or 400 FC/100 ml. The NTAC determined that one-half of that number, 200 FC/100 ml, would be an appropriate threshold density for health risk indication. Despite deficiencies in the data upon which the 200 number was based, no other criteria were recommended. Therefore, most states adopted it as a standard by 1968.

The degree to which indicator organisms represent the presence of pathogens has been the subject of continuing investigation (Thomann 1987). There is a general correlation between the concentration of FC bacteria and the occurrence of *Salmonella* (Geldrich 1978). In general, at levels of FC less than 200 FC/100 ml, *Salmonella* occurrences were less than about 30%. However, at levels of FC at 1,000 FC/100ml or greater, occurrences of *Salmonella* range to greater than 95 percent.

In 1972, a number of studies were initiated by the EPA for the purpose of developing a new set of bacterial criteria. By 1986, the EPA adopted revised indicator bacteria, namely *Escherichia coli (E. coli).* This is a bacteria of the fecal coliform group and enterococci, a subgroup of the fecal streptococcus group. These bacteria were stipulated in a 1986 EPA publication, "Ambient Water Quality Criteria for Bacteria." The EPA studies concluded that *E. coli* and enterococci had the best correlation with gastroenteritis rates at bathing beaches. (The authors have encountered some difficulty in using *E. coli* for receiving-water quality evaluations for several CSO studies. Mass balance modeling at these problem locations suggests that *E. coli* discharges may not be amenable to traditional mass balance modeling approaches. The cause for these problems is uncertain but may be due to the differential effects of measurement methods between CSO and receiving-water samples.) Still, several states have adopted the new criteria and are applying them to CSO situations.

The national CSO control policy and some states recognize the highly uncertain applicability of even the newest bacteria standards during wet-weather conditions and have considered this in their applications. As one example, the North Carolina bathing-beach standards specifically state that the criteria are *not* applicable during or immediately following periods of rainfall. The state of Maine's standards specifically base bacterial criteria in bathing waters on *E. coli* of human origin, with most of the criteria applying only for the period from May 15 to September 30. During other times of the year, the criteria call for a bacteria content limit that "shall be as naturally occurs." Virginia has elected to retain fecal coliform standards but flags only exceedances of the standards when both the geometric mean *and* the maximum standard are consistently exceeded.

The ratio of FC to fecal streptococcus (FC/FS) has been used as a rough indicator of whether the sources of bacteria are from humans (FC/FS > 4) or from other warm-blooded animals (FC/FS < 1) (Geldrich 1966). The use of the FC/FS ratio appears to represent one of the only methods, short of direct pathogen measurement, of discerning the difference between human and animal origin. In 1988, several beach closings along the eastern coastal waters of New York and New Jersey brought to the attention of regulatory agencies and municipalities the importance of having a reliable means of protecting the public health partly due to the fact that storm water can cause violations of the FC, *E. coli,* and enterococcus standards. However, there is little documentation ascribing waterborne human illness to bacteria of animal origin. In the absence of a more reliable standard, the FC/FS ratio continues to be used. Ratios of the more specific *E. coli* (part of the FC group) to enterococcus (part of FS) may eventually be established and provide a more reliable measure of bacterial origin. It should be noted that due to the varying decay ratios of FC and FS, the FC/FS ratio should be used carefully and within certain limitations, as outlined by Thomann (1987). In general, it may serve as a broad indicator of origin and signal the absence of human bacteria and the persistence of less harmful nonhuman bacteria often associated with storm water. Caffeine has proven effective in the Seattle, Washington, area as a tracer for human wastewater and may be used elsewhere in con-

bacterial measurements to assess the degree of human health risk (King County Metropolitan Services 1996).

Aesthetic Deterioration

Aside from public-health and aquatic-health risks, CSOs can cause serious impairment of aesthetics in receiving waters. CSOs carry an assortment of obvious and unsightly debris from runoff and wastes from sanitary sewage. Increased suspended solids, sanitary debris, plastics, floatables, oils, and grease are all common complaints related to CSO. Such objections are common to all receiving waters but may persist for different periods depending on the current strength and dilution. Although aesthetic impairment may not reduce aquatic health or cause obvious human health risk, it does have socioeconomic significance. CSO debris and suspended solids can reduce recreational use and appreciation of receiving waters, thereby limiting recreational expenditures and near-shore development. Increased suspended solids can cause reduced water clarity affecting recreational use. Quantitative measures of aesthetic reduction are rare but possible (AMSA 1996). Visual observations, postaudits of voluntary annual cleanups, and resident interviews can be used to define the scope of the problem related to CSO and perhaps even be used as measures of CSO control performance (AMSA January 1996).

Monitoring of Impacts

In every CSO study, there is a recurring debate on the best means to define CSO impacts and benefits of alternative control strategies. Should it be monitoring or modeling? Practically speaking, it is not an either/or decision but rather how much of each. The two activities need to be balanced and designed for an integrated assessment. Monitoring is needed to identify existing problems, define baseline conditions, and support the development of reliable and verified mathematical models. Nothing provides stronger evidence of existing receiving-water impacts than actual measurements and observations, but it is difficult to extend them to provide meaningful forecasts of benefits expected from CSO controls. All CSO evaluation studies should include measurements and observations of post-storm receiving-water impacts caused by CSOs. However, monitoring alone has limitations because the monitored conditions:

1. may be unrepresentative;
2. may not capture infrequent but severe events;

3. do not delineate the separate effects of CSO and other nonpoint and point source pollution; and
4. cannot be used to compare proposed CSO controls.

Therefore, modeling should generally be used to complement monitoring and to forecast and analyze conditions other than those monitored. On the other hand, models need to be supported by and calibrated to actual data. Monitoring and modeling must therefore be designed together as an integrated program with full consideration of the types and extent of data needed to define the problems. (More will be provided on modeling in later sections of this chapter.)

Monitoring programs should include initial wet-weather reconnaissance surveys involving a few stations downstream of CSOs and at least one station upstream. The reconnaissance monitoring should be conducted during seasons expected to encompass the most significant impacts and should extend from a day before a storm to as much as four days afterward. Parameters should include BOD, dissolved oxygen, temperature, bacteria, and other parameters suspected as having impacts, plus visual observations for debris and floatables. Occasionally, routine fixed-frequency-monitoring data can be used to assess trends, the presence of impacts, and the periods or seasons when the receiving water is most susceptible to CSO discharges. Most fixed-frequency samplings are too infrequent to adequately characterize storm impacts and recovery. Therefore, surveys must be designed with this specific need in mind.

The reconnaissance surveys should be used as a basis to design more intensive surveys in terms of sampling location, duration, and frequency. Subsequent to the reconnaissance survey(s), at least two or three additional intensive evaluations are recommended to better define impacts for different conditions and support model development. Often, impacts are evident only for large storms and not apparent for the smaller events. Hence, multiple appraisals are recommended for defining the recurrence of problems and representing a fuller range of conditions.

In general, station locations and frequency for intensive assessments must be in sufficient numbers to clearly define spatial and temporal trends. This usually requires sampling every 3–6 hours at stations located with time of travel between locations of approximately 6–12 hours. In rivers, the sampling logistics are the simplest because all impacts flow downstream. Station locations and timing are easy to define. For estuaries, choosing station locations is still relatively easy, but consideration of tidal movement and timing becomes important. Samplings at fixed times in each tidal cycle (usually a slack time) are most desired to help standardize analysis and presentation of the data. For lakes, timing is less difficult, but station locations are more difficult to select due to multidimensional and sometimes erratic transport. In lakes, drogue or current studies can be used to define current transport. The circulation patterns in a lake may vary depending on wind stratification and other variable factors. Therefore, a network of lake stations should be set up to encompass the probable horizontal and vertical transport directions. Preliminary data can be used to refine the sampling network. Also, reconnaissance data may show minimal localized effects, in which case coarse whole-lake sampling may be more appropriate than a localized network. In all cases for rivers, estuaries, and lakes, the reconnaissance sampling is indispensable for defining adequate station locations and sampling frequency.

The measurements should focus on the anticipated CSO-related problems. However, CSO impact surveys should routinely include measurement of dissolved oxygen, BOD, ammonia, and bacteria. Temperature and visual observations of cloud cover, sunlight, water transparency, floatables, debris, and other conditions are inexpensive and should also be routine. Daily composites are usually suitable for other parameters such as nutrients and chlorophyll. Toxic substances, such as trace metals, typically need be measured only immediately before the storm to define prestorm conditions, immediately after the storm to define maximum impacts, and at the end of the survey to determine persistence. Diurnal dissolved oxygen variations due to photosynthetic activity can be very important and, if present, generally dictate the need for continuous monitoring and/or station measurements at three-to-five-hour intervals. Photosynthetic activity can often confuse or mask the interpretation of CSO impacts on dissolved oxygen, especially since a reduction in both sunlight and photosynthesis usually accompanies storms. Therefore, if high productivity is present, careful evaluation of diurnal and light conditions is essential.

Special note should be made concerning the scheduling of CSO impact monitoring. Obviously, it is impossible to plan for storms to occur convenient for sampling. It is also generally impractical to use automatic samplers for the receiving water. It is therefore very important to have a well-developed mobilization plan in order to capture immediate poststorm impacts. Impacts related to solids, bacteria, and toxics are best captured immediately after the storm, during the first flush of pollutants from the sewers. A rapid mobilization is essential. The use of weather forecasting services is often a valuable aid. Also, routine fixed-frequency monitoring programs can be a valuable source of prestorm data and for identifying cumulative impacts.

With data in hand, the planner/engineer can proceed to analyze more fully a wider range of conditions after developing and calibrating a mathematical model. Models are well suited for providing information not available from data alone. In particular, models can be used to simulate water quality for conditions not monitored. This could include larger, more intense, storms, or more critical environmental conditions such as lower stream flow. It is generally infeasible to extrapolate to nonmonitored conditions from data due to the large number of variable factors. The power of modeling is that virtually all such conditions can be assessed quickly. Furthermore, models can be used to examine impacts and benefits from proposed CSO control strategies.

In summary, monitoring is essential to all CSO studies to demonstrate and define the presence of CSO impacts. Modeling provides a complementary tool to analyze impacts at nonmonitored conditions. However, models must be based on and confirmed with actual physical, chemical, and biological information. Therefore, monitoring and modeling must be designed and conducted hand-in-hand.

Modeling Approaches

Mathematical models are a powerful tool for assessing CSO impacts and for comparing CSO control alternatives. However, the effective use of receiving-water modeling requires the selection of an appropriate framework and the site-specific validation with data. General guidance is provided in this chapter for the selection and development of suitable models for CSO assessments. Mathematical models use algebraic and differential equations to simulate the physical, chemical, and biological processes that determine water quality. Modeling consists of three basic components:

1. Model Formulation
   a. system equations
   b. equation solutions
2. Model Validation
   a. model inputs
   b. comparison to data
   c. coefficient adjustment (calibration)
3. Model Use
   a. selection of forecasting conditions
   b. comparisons of results

The model formulae selected must be appropriate for the site and problem. Obviously, equations describing channel flow are not well suited for lakes. Other decisions are not always so obvious but are equally important depending on the complexity. The solution to these equations can either be desktop analytical or involve a computer program. Neither inherently offers more reliability, but computer models often offer more flexibility. Provided an appropriate model framework has been chosen, the inputs such as flow, pollutant loads, temperature, and reaction rates must still be carefully selected to characterize the site properly. Then, before use in problem solving, the model reliability should be validated by comparison to data. Overall, despite the apparent burden of these steps, models are essential tools in analyzing conditions and alternatives not monitored. Only through the use of model simulation can the engineer/planner compare the merits of different CSO control strategies.

Helpful Concepts

The first question usually asked of models is: "What is the impact of CSO on water quality?" Fortunately, water-quality models are very well suited for analyzing the incremental impacts from various sources. The nature of the equations used in water-quality models allows the application of the concept of superposition. This mathematical concept permits each individual source of pollution impact to be modeled separately. The total water-quality condition is then the sum of component impacts simulated separately. Shown in Figure 3–4 is an illustrative example of this component analysis for a river impacted by background, upstream, separate storm water, and CSO impacts.

Superposition is a powerful technique that allows the modeler easy flexibility in answering important questions. For example, CSO impacts can be incrementally compared to other nonpoint wet-weather impacts or to dry-weather baseline water quality. Establishing the relative importance of CSO impacts to other sources can be very important in developing cost-effective, basin-wide pollution solutions, and establishing the need for and benefit from CSO controls.

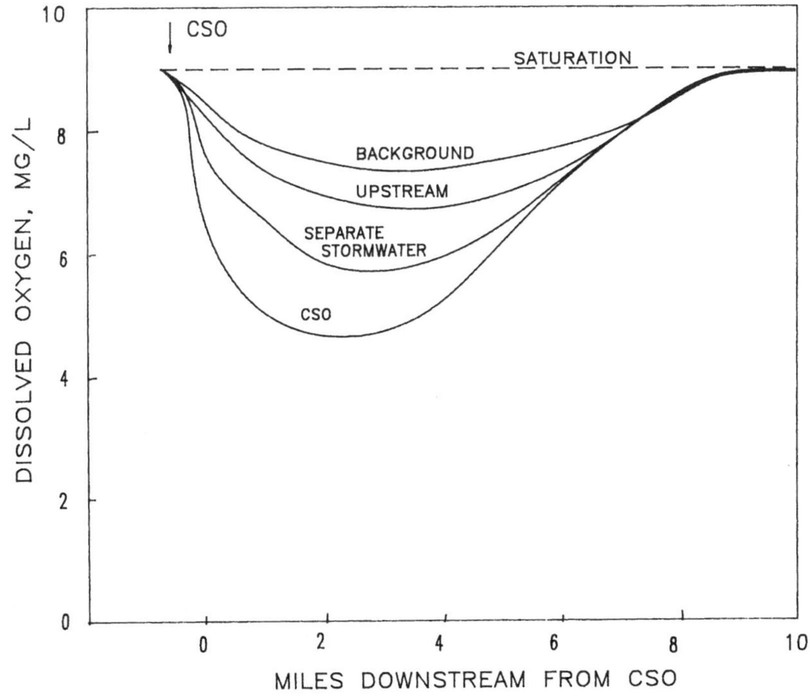

Figure 3–4. Component analysis of river impacts on dissolved oxygen.

Often, other pollution sources can be more important than CSO, but monitoring alone does not establish their relative impacts. Superposition can also be used to compare individual CSO effects to help devise cost-effective prioritization of controls. Last, superposition can also be used to analyze model performance and sensitivity.

Analyzing CSO impacts separately is an important conceptual approach. CSO-derived pollutants are often more reactive than other receiving-water pollutants because CSOs include a component of raw sewage. As a result, model simulation of CSO impacts often requires reaction rate coefficients different from those required by simulation of impacts from treated discharges. In addition, CSO control programs are often based on an evaluation of cost-effective incremental improvements, not absolute compliance to sometimes inappropriate "dry weather" water-quality standards. This approach is often preferred because wet-weather conditions are too variable, factors other than CSO often control compliance, and many water-quality standards are not appropriate for wet weather.

A common dilemma among new modelers is how to conveniently analyze the transient nature of CSO impacts. Often the first approach is to utilize complex models that can characterize all the transient aspects of the system and storm under study. However, two simpler approaches are available. One can always assume the loading and related conditions are constant or steady-state. This provides a conservative estimate of the impact but often is representative if loading points are close together and dispersional mixing is small. Analyzing the problem in a moving reference (Lagrangian) framework is another approach. This involves simulating the dynamics of a plug of polluted waters as it flows or circulates downstream, ignoring the dynamics and surrounding system. This method is very useful if CSOs are either quite close to each other and can be analyzed together or are far enough apart to be analyzed separately.

Having provided some overall conceptual guidance on modeling approaches, let us examine some specifics related to different types of receiving-water bodies. Again, the objective of this chapter is general guidance, not textbook instruction on modeling. For more thorough technical guidance, the reader is referred to other resources (Thomann 1972; Thomann and Mueller 1987).

Streams and Rivers

The modeling mechanics for streams and rivers are the simplest because the dilutionary flow is well quantified and the direction of flow, obvious. If one assumes that the CSO pollutant inputs are well mixed across the stream/river channels, then the characteristic equations are simple. The well-known Streeter-Phelps equation for dissolved oxygen is a good example. This classic equation (and related dilution equations) can be used to simply characterize water-quality impacts in many streams and rivers. The basic equations are:

$$C_0 = W/Q$$

$$C = C_0 \exp(-Kt)$$

$$D = D_0 \exp(-K_a t) - L_0\left(\frac{K_d}{K_a - K_r}\right)[\exp(-K_r t) - \exp(-K_a t)]$$

where:

C_0 = initial pollutant concentration (M/L^3)

C = pollutant concentration downstream (M/L^3)

W = total pollutant loading (M/T)

Q = total river flow (L^3/T)

t = time of passage (T)

K = pollutant decay rate ($1/T$)

D = dissolved oxygen deficit downstream (M/L^3)

D_0 = initial dissolved oxygen deficit (M/L^3)

L_0 = initial BOD concentration (M/L^3)

K_a = atmospheric reaeration rate ($1/T$)

K_d = BOD deoxygenation rate ($1/T$)

K_r = BOD loss rate ($1/T$)

By employing the concept of superposition, one can use these basic equations quickly and simply to assess CSO impacts in many streams and rivers. Additional pollutant sources or components can be analyzed separately and are added to baseline conditions to simulate overall water quality.

The apparent complexities in CSO analysis, due to transient CSO loading, are relatively easily handled in streams and rivers. The modeler can either conservatively assume that all the loads are steady and compound or instead focus attention on only one plug of water moving downstream. This allows an assumption of constant loading (averaged for the plug) and then a tracking of this plug as it goes downstream. Higher and lower loading rates from a CSO can be analyzed separately since the pollutants travel in a train with limited intermixing. Additional compounded impacts from downstream loads can be added through superposition by separately tracking the time of passage with the time of the loadings.

The preceding equations are obviously very basic, and the reader is again referred to other standard modeling textbooks for more details and added complexities (Thomann and Mueller 1987). Nonetheless, these basic equations and approaches are common to all modeling of CSO impacts. Various computer models are available that utilize these equations and concepts as a basis but offer added flexibility to handle complexities in reactions, loadings flow, and geometry. These models include examples such as PULSEQUAL (LTI 1992a, b), WASP (EPA 1988), AUTO-QD (EPA 1973), QUAL2E (EPA 1987), and RECEIVE (Raytheon 1975) as summarized in Table 3–1. For selection and use of these models, the reader is referred to the corresponding manuals.

Estuaries

CSO impacts on estuaries can often be addressed in a manner similar to that of stream systems but with additional complexity due to dispersion, tidal hydraulics, and saline intrusion. If the upper freshwater portion of an estuary is being examined, then simple one-dimensional, intratidal analyses can be appropriate. If assessments require examination of intertidal, short-term impacts or involve the lower, more lakelike portion of the estuary (where stratification due to saline intrusion is a major process), then more complex two- and three-dimensional dynamic computer models may be needed.

Impacts of CSO discharge in the upper estuary can often be analyzed as a slug of impacted water, as for streams and rivers. However, in addition to the advective transport in rivers, the slug is also affected in estuaries by tidal motion and dispersional mixing. This tends to distribute and dilute the slug over time. This effect of dispersion is displayed in Figure 3–5. In addition, since tidal action moves the slug back and forth as it progresses downstream, modeling must either account for this motion or be calculated at consistent points in the repeating tidal cycle. Surveys used for model development are often conducted at the points of low or high slack tides to allow comparison to intratidal modeling results.

As in streams and rivers, there are analytical, closed-form equations to simulate the transport, mixing, and reaction of pollutants in estuaries. The calculations can conservatively assume the loadings are steady or assume the loadings occur as an instantaneous slug. In both cases, the equations can calculate the transport downstream, the mixing due to tides, and the reaction gains or losses.

The slug or pulse loading equation is usually preferred for analyzing CSO impacts because the steady-state equations underestimate the dilution effects from tidal mixing. Assuming full lateral and vertical mixing, the basic water-quality equations to simulate the impacts of pulse loading are:

$$C = SM\exp(-Kt)$$
$$D = SM_L(K_d/K_a - K_i)[\exp(-K_rt) - \exp(-K_at)] + SM_D\exp(-K_at)$$
$$S = \exp[-(x - ut)^2/4Et]\big/[2A(\pi Et)^{1/2}]$$

Table 3–1. Summary of Selected Stream and River Computer Models for CSO Impact Assessments

	Transport		Geometry			Variability		Parameters			
Model	Dimensionality	Dispersion	Branching	Channels	Multi-segment	Upstream	CSO	Dissolved Oxygen	Algae	Toxics	Bacteria
PULSEQUAL	1	Y	N	Rect	Y	Y	Y	Y	N	Y*	Y
DOSAG-3	1	N	Y	Rect	Y	N	N	Y	N	Y*	Y
AUTO-QD	1	Y	N	Rect	Y	Y†	Y†	Y	N	Y	Y
RUNQUAL	1	Y	Y	Trap	Y	Y	Y	Y	N	Y	Y
WRECEV	2‡	Y	Y	Rect	Y	Y	Y	Y	Y	Y*	Y
HSPF	1	Y	Y	Rect	Y	Y	Y	Y	Y	Y*	Y*
QUAL-I	1	Y	Y	Rect	Y	N	N	Y	N	N	Y*
QUAL-II	1	Y	Y	Rect	Y	N	N	Y	Y	Y*	N
QUAL-III	1	Y	Y	Rect	Y	Y†	Y†	Y	Y	Y*	N

*Simulate by using BOD and BOD kinetics as a surrogate.

†Quasi-dynamic using multiple steady states.

‡Quasi-two-dimensional setup available using link-node network.

Y = Yes

N = No

Rect = Rectangular

Trap = Trapezoidal

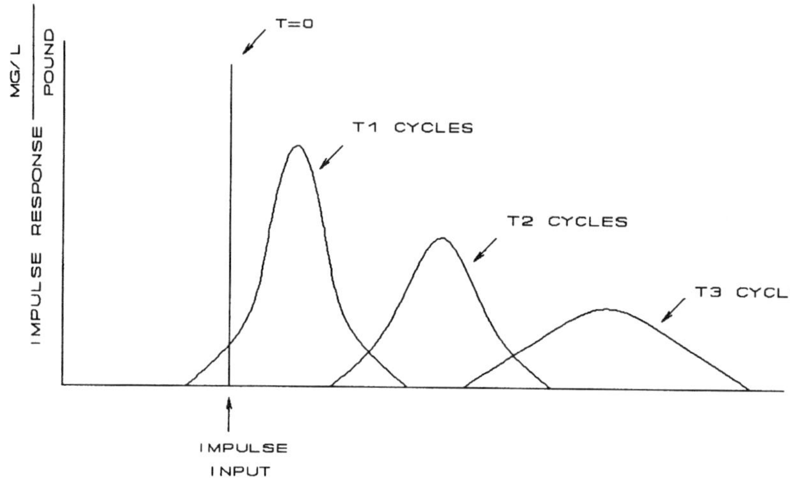

Figure 3–5. Effect of dispersion on instantaneous discharge (Thomann 1972).

where:

A = area (L^2)

E = longitudinal dispersion (L^2/T)

x = distance downstream (L)

M = mass loading of a pollutant (M)

M_L = BOD mass load (M)

M_D = dissolved oxygen deficit mass load (M)

U = average velocity of stream (L/T)

and the other terms are as applied before. Again, other sources or components (such as NBOD) can be added by superposition.

The approach just shown allows calculations of CSO impacts in estuaries on a tidal-averaged basis. This approach generally requires the modeler to sum nearby loads and ignore small-scale differences due to intertidal processes. Rapid and large tidal mixing generally lends greater validity to these assumptions. Therefore, these simple equations provide powerful methods for screening alternatives and comparing CSO impacts.

Although the desktop formula approaches are highly recommended for many situations, site and problem characteristics can often dictate the need for a more rigorous and complex analysis. Widely distant, varied, and oddly timed CSO loads coupled with highly variable estuary conditions are examples of factors limiting the applicability of the simpler analysis. In these cases, the modeler must generally use one of several computer models specifically adapted for estuaries—WASP (EPA 1988), PULSEQUAL (LTI 1992; 1993b), AUTO-QD (EPA 1973), and the Dynamic Estuary Model (EPA 1979) are examples of one-dimensional, tidal-averaged, and intertidal models. For estuary sites with multidimensional transport, the level of model complexity is beyond the scope of this chapter. The reader is referred to one of several available models highlighted in Table 3–2.

Lake Analysis

Analyzing the impacts of CSOs on lakes and impoundments is much different from and generally more complex than for rivers or estuaries. First, and most obvious, the flow and circulation in lakes is much more varied and complex. Second, the nature of water-quality problems and processes is often different in lakes. As a result, lake analysis usually requires specialized computer modeling programs and comprehensive monitoring programs. The CSO impacts in lakes can generally be categorized into three groupings: immediate end-of-pipe impacts, whole-lake impacts, and spatially patterned impacts. End-of-pipe impacts and whole-lake impacts can often be analyzed using simple analytical equations, but spatially patterned impacts almost always require a complex computer model.

End-of-pipe impacts are those in the immediate discharge plume from a CSO and usually involve toxicity concerns. The research on plume mixing and dilution is extensive, and formulae are available to calculate the impacts depending on the physical characteristics of the lake outfall site (Fischer et al. 1979; EPA 1991). Factors to consider when selecting an individual formula include:

- Near-field versus far-field
- Outfall submergence
- Discharge buoyancy
- Discharge velocity
- Shore and bottom effects

The formulae for ambient mixing are typically simple to apply, but mixing induced by the turbulence of a discharge jet is more complex. Proper selection of the formulae or mixing zone model is essential for reliable calculations. (See Mixing Zone section for more detail).

Whole-lake impacts are very different and consider the accumulation of pollutants from CSO on a lake-wide basis. Again, research in this area is extensive, and different formulae are available to address various types of problems for lakes with different characteristics (Chapra and Reckhow 1983). In general, the simpler equations are based on a completely mixed mass balance characterization of a lake with flow in and flow out. To calculate lake transients from a slug

Table 3–2. Summary of Selected Estuary Computer Models for CSO Impact Assessments

		Transport			Geometry			Variability		Parameters			
Model	Dimen-sionality	Advective	Dispersive	Tidal	Branching	Channel	Segmen-tation	Upstream	CSO	Dissolved Oxygen	Algae	Toxics	Bacteria
PULSEQUAL	1	Y	Y	Y	N	Rect	None	Y	Y	Y	N	Y†	Y
WASP4	3	Y	Y	Y	Y	Rect	Box	Y	Y	Y	Y	Y	Y
WRECEV	2*	Y	Y	Y	Y	Rect	Link-Node	Y	Y	Y	Y	Y†	Y
EXPLORE-I	2*	Y	Y	Y	Y	Rect	Link-Node	N	Y	Y	Y	Y	Y
DEM	2*	Y	Y	Y	Y	Rect	Link-Node	N	Y	Y	Y	Y	Y
CAFE/DISPER	2	Y‡	Y‡	Y‡	N‡	—	Plume	N	Y	N§	N	Y	Y
CEQUAL-W2	2	Y	Y	Y	Y	Rect	Box	Y	Y	Y	Y	N†	Y

*Quasi-two-dimensional setup available using horizontal link-node networking.

†Simulate nonconservative using BOD and BOD kinetics as a surrogate.

‡Plume calculations in advective/tidal flow field with dispersion.

§Deoxygenative material concentrations can be calculated, but not Dissolved Oxygen concentrations.

Y = Yes

N = No

Rect = Rectangular

Trap = Trapezoidal

of pollutants completely mixed in the lake, the basic equation is:

$$C = C_0 \exp[(-K - Q/V)t]$$
$$C_0 = M/V$$

where:

Q = flow (L^3/T)

V = lake volume (L^3)

and other terms are as defined previously. For a steady-state constant load calculation, the lake equation is:

$$C_{ss} = \frac{W}{(Q + KV)}$$

where:

C_{ss} = steady-state pollutant concentration (M/L^3)

However, both of these equations and their respective assumptions are very simplistic. They are usually suited only for performing a first-screening analysis to define the potential of CSO impacts in lakes. The first approach based on effluent mixing alone usually overestimates localized impacts, whereas the second approach usually underestimates the impacts. Nonetheless, the calculations are valuable for developing a quantitative "feel" for the lake response to CSO inputs.

In contrast to end-of-pipe and whole-lake effects, spatially patterned impacts are much harder to analyze. These are transient impacts that originate from CSOs but circulate, disperse, and dissipate as waters move the impacts to different lake locations. Mixing zone models (discussed in the next section) can be applied to assess near-field effects if the general circulation pattern is known in the vicinity of the outfall. To characterize these impacts fully, complex computer models are usually needed, several of which are referenced in Table 3–3. These models rely on the same principles of mass conservation as the simpler models but usually are constructed to represent the lake as a 2- or 3-dimensional network of mass balance calculation cells. Mass balance equations are written for each cell and then solved simultaneously by a computer program.

Mixing Zones

Mixing zone models are mathematical tools used to predict the rates at which pollutants are initially diluted by dispersion and transport in the river. Different model parameters and coefficients need to be defined before using these models. The parameters are different from one model to another, based on the conditions

Table 3–3. Summary of Selected Lake Computer Models for CSO Impact Assessments

			Transport		Variability			Parameters			
Model	Dimensionality	Plane	Advective	Dispersive	Tidal	Background	CSO	Dissolved Oxygen	Eutrophication	Toxics	Bacteria
WASP4	3	—	Y	Y	Y	Y	Y	Y	Y	Y	Y
LAKSCI	1	Vertical	Y	Y	N	Y	Y	N	Y	Y	Y
EPARES	1	Vertical	Y	Y	N	Y	Y	Y	N	Y	Y
HSPF	1	Horiz/Vert	N	N	N	Y	Y	Y	Y	N	N
CAFE/DISPER	2	—	Y	Y	Y	Y	Y	N	N	Y	Y
PLUME	2	Horiz/Vert	N	Y	N	N	N	N	N	Y*	N
LAKECO	1	Vertical	Y	Y	N	Y	Y	Y	Y	Y	N
WQRRS	1	Horizontal	Y	N	N	N	Y	Y	Y	N	N

*Simulate by using BOD and BOD kinetics as a surrogate.

Y = Yes

N = No

Horiz = Horizontal

Vert = Vertical

of the modeled area and the theory behind the model. Generally speaking, ambient conditions of a water body are expressed in terms of geometric parameters such as the shape of the water body, vertical cross-sections, and bathymetry near the point of discharge, in addition to hydraulic parameters like velocity and density distributions. Discharge characteristics are given in terms of flow rate and geometric and transport flux characteristics.

Two regions can be defined when dealing with mixing zones: "near-field" and "far-field." The near-field refers to the region close to the outfall within zero to approximately hundreds of feet where the initial jet characteristics of moment flux, buoyancy flux, and outfall geometry influence the jet trajectory and mixing more than the ambient water conditions (Figure 3–6). Here, acute aquatic toxicity criteria are applied for evaluation of pollutant impact. Far-field refers to the region far from the outfall where such effects are dissipated and ambient conditions predominantly influence the trajectory and mixing of the plume. At this point, chronic aquatic toxicity is usually examined. Considerations of criteria application are discussed in agency guidelines (EPA March 1991). Due to their episodic nature, most CSO evaluations focus on acute mixing zones in the near-field zone within approximately one hour of the discharge.

The simplest approaches for estimating CSO dilution at the edges of the near-field and far-field plumes are single equations derived using conservative mixing assumptions. A minimum estimate of very near-field CSO dilution can be made with the following jet momentum equation (EPA 1991):

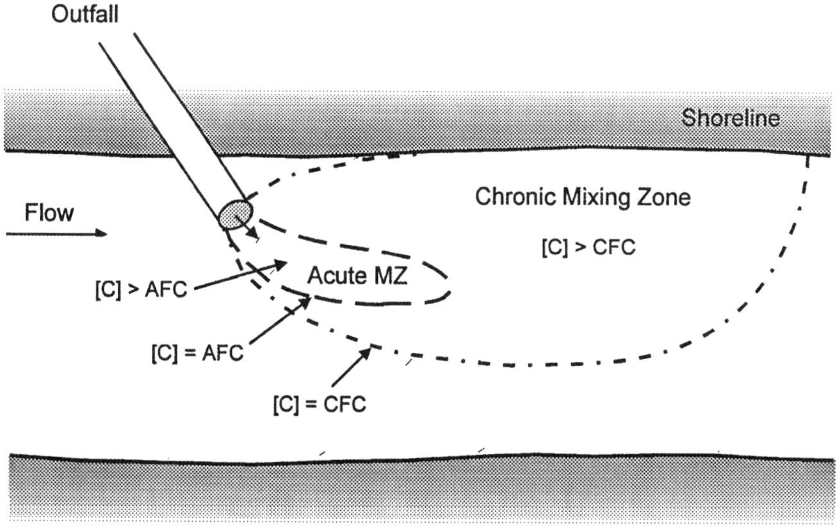

Figure 3–6. Illustrative mixing zone and application of acute and chronic criteria.

$$S = 0.3\ (x/d)$$

where:

S = minimum dilution (dimensionless)

x = distance to edge of plume (L)

d = diameter of the outfall (L)

This equation assumes that the ambient mixing (mixing caused by the receiving-water turbulence) is zero and the effluent buoyancy is neutral. As the effluent moves farther from the outfall, the more this equation diverges from reality due to the increasing influence of ambient mixing.

Far-field ambient mixing can also be estimated for neutrally buoyant discharges to rivers using simple equations (EPA 1991):

$$S = Q_r{*}(\pi{*}D_y\,x/u)^{1/2}/(Q_e{*}W)$$

$$D_y = 0.6y(gys)^{1/2}\ \pi 50\%$$

where:

Q_r = river flow (L^3/T)

Q_e = effluent discharge (L^3/T)

D_y = lateral dispersion (L^2/T)

x = distance from outlet (L)

u = instream water velocity (L/T)

W = river width (L)

y = water depth (L)

g = acceleration of gravity (L/T^2)

s = channel slope (dimensionless)

These ambient mixing equations apply far-field, outside the near-field area that is primarily influenced by discharge-induced mixing. As indicated, lateral mixing (D_y) can vary by 50% or so depending directly on the sinuosity ("curvedness") of the river (more bends in the river creates more mixing).

Several mixing zone models are available to evaluate more complex CSO discharge situations under conditions of localized steady-state (e.g., minimal variability within a short distance from the outfall). Each mixing zone model is appropriate to a specific set of discharge and receiving-water situations. Table 3–4 summarizes the important characteristics of the available models and can be used for matching a model to particular study needs. For example, the PDS model would be appropriate for estimating surface CSO discharges to deep river or estuary systems.

Table 3-4. Summary of Selected Mixing Zone Computer Models for CSO Impact Assessments

Feature	Jet Momentum	River Initial Mix	Ambient Dilution	CORMIX	UM-PLUMES	RSB	PDS	PDSM	UDKHDEN
Discharge	Submerged	Either	Either	Either	Submerged	Submerged	Surface	Surface	Submerged
Receiving water	Any	Shallow	Shallow	Any	Any	Deep	Deep	Deep	Any
No. of ports	Single	Single	Any	Any	Any	Multiple	Single	Single	Multiple
Dimensions	One	Zero	One	Three	Three	Three	Three	Three	Three
Complexity	Low	Low	Low	Moderate	Moderate	Moderate	High	High	High

Model ambient river parameters for the models are extensive and may include width, water depth, water temperature, ambient velocity, angle of the outfall, effluent velocity, salinity, heat conduction coefficient, horizontal and vertical dispersion coefficients, jet entrainment coefficient, and water temperature. The entrainment coefficient expresses the shearing action applied by ambient water against the plume as the jet cuts through the water current. Horizontal and vertical dispersion coefficients define the spread of the plume in two directions. Salinity, temperature gradient (between discharged and ambient water), and heat conduction coefficient control the buoyant characteristics of the plume.

An application of the CORMIX model is shown in Figure 3–7 demonstrating the usefulness of assessing minimum dilution moving outward from the outfall in the Willamette River.

Sediments

The impact of CSOs on river, estuary, and lake sediments is a common concern but is difficult to analyze. Unlike CSO impacts on waters in rivers, estuaries, or

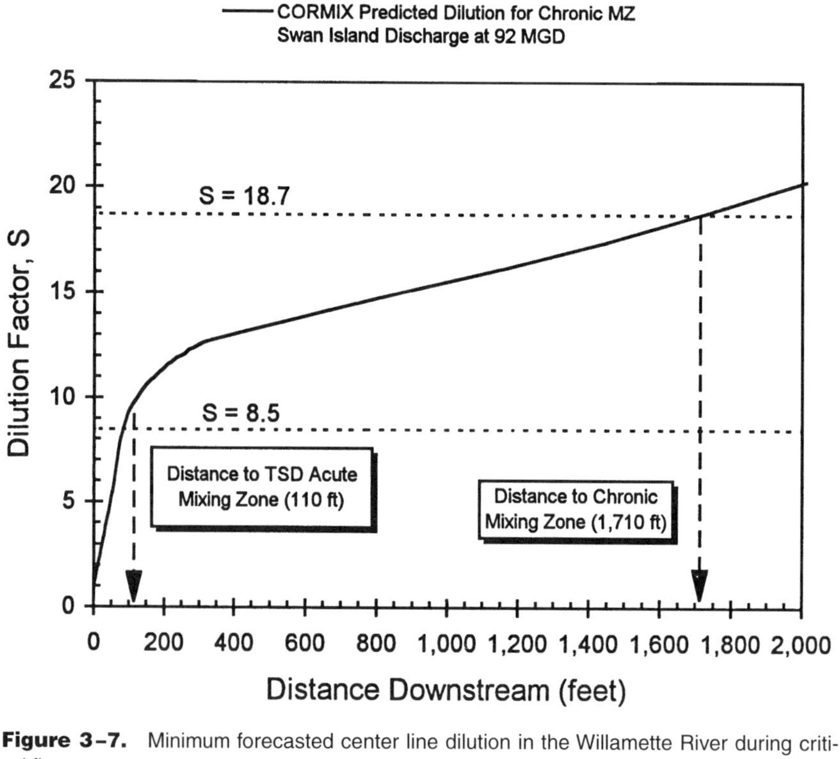

Figure 3–7. Minimum forecasted center line dilution in the Willamette River during critical flow.

lakes, there are no simple calculations that can be performed to assess these impacts for sediments. Unfortunately, comprehensive sampling and modeling are typically required. The potential of a problem can often be determined by an examination of pollutant concentrations in CSO solids and by a comparison of CSO solids loads to loads from other sources. However, confirmation of the problem and accurate forecasting of the benefits from CSO controls require more rigorous analysis. Discussion of these considerations is beyond the scope of this chapter.

Critical Design Conditions

In using models to plan CSO controls and protect water quality, the issue of critical design conditions always arises. Engineers and planners want to size and select controls as necessary to protect water quality for all conditions equal to or less critical than a specified critical condition. This concept evolved from similar planning needs for wastewater-treatment-plant effluent limits. In this case, limits are usually determined to protect water quality at conditions equal to or less critical than a specified drought flow. Hence, by comparison, there is a similar desire to establish design conditions for CSO control. Unfortunately, the parameters and conditions are more difficult to define because of the wide variability.

The impact of CSO on receiving waters depends on dilution, dictated by flow; reaction rates, dictated by temperature; and CSO loading rate and volume, dictated by rainfall. Selecting a combination of these conditions as a critical design condition is difficult because the three do not ordinarily vary together in a predictable fashion. High CSOs from large rains typically do not coincide with low drought flows. As a result, a variety of combinations can be critical, ranging from moderate rains at drought flow to torrential rains at moderate flows. Hence, selecting one critical condition is usually inappropriate and often impractical. Nonetheless, evaluating CSO impacts and benefits of proposed controls requires that water quality be identified along with associated abatement costs. Two general approaches are feasible for these evaluations. First, a group of design conditions can be used to represent the range of expected impacts. Second, a range of conditions can be analyzed using historical data, and then benefits and impacts can be characterized using a frequency and severity assessment.

Selected Conditions

Evaluating CSO impacts for individual storms is generally the easiest approach to conduct and understand. However, for the information to be useful, a number of different storms must be examined at varying environmental conditions (e.g., flow, temperature, and nonpoint loadings). A set of impacts or benefits can then be used for quantitative comparisons employing these varied conditions. The engineer/planner can then analyze the impacts and effectiveness for controls over a range of conditions that would be representative of the expected impacts during a typical year.

The simplest approach for selecting design conditions is to evaluate the rainfall history statistically and to select rainfalls with different occurrence frequencies. Examples might be water-quality calculations for a once-a-month storm and twice-per-month storm during typical summer river conditions. Shown in Figure 3–8 is an example of this approach for Richmond, Virginia. However, even with

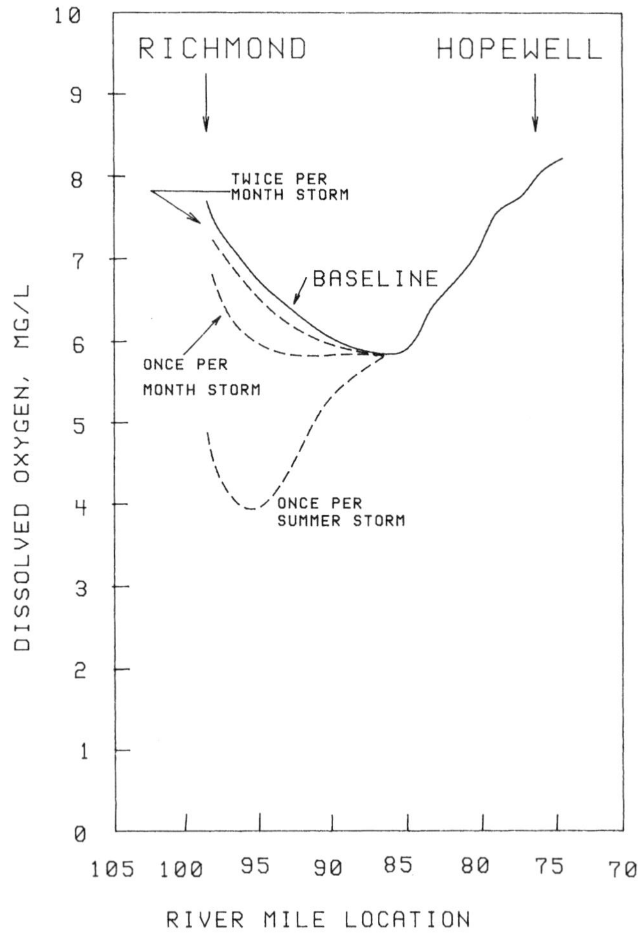

Figure 3–8. Three summer-design-storm dissolved-oxygen simulations (flow = 2,000 cfs; day 3, temp. = 29.5°C) for the upper James River Estuary (Limno-Tech, Inc., from Greeley and Hansen et al. 1987).

this method, the selection of companion river flow, temperature, and other conditions is difficult to statistically characterize or select. This can be very significant if forecasting compliance to water-quality standards is an objective. Choosing a flow too small or too large can produce unrepresentative results.

As indicated, the preceding approach is easy to conduct and understand but gives limited information. In the example given, three conditions were examined. More conditions over a wider range would be better. However, the collective frequency of problems is still difficult to forecast because of parameter variations. The selection of companion flow, temperature, and other non-CSO loads to correspond to each storm offers a dilemma; the impacts will obviously differ depending on these selections. A solution to this problem is to use environmental conditions that actually correspond to the selected storms or to use statistically selected companion conditions. Nonetheless, the selection is still difficult, and the result, limited.

Frequency Analysis

A better, but more involved, approach is to analyze water quality under the whole range of conditions and to determine the joint probability of rainfall and water quality impacts. Results can then be presented in terms of frequency, severity, and duration of impacts. This analysis can be conducted either by running continuous simulation for a five-to-ten-year history of storms and river conditions or by probabilistic simulations (e.g., Monte Carlo). In continuous simulation, a long history of storms is matched to actual environmental conditions, and each storm is analyzed individually or continuously. Shown in Figure 3–9 are example results of continuous simulations for Richmond, Virginia, and South Bend, Indiana.

A second approach is to run repetitive simulations using statistically selected inputs for rain, flow, temperature, and so forth, and applying a Monte Carlo probabilistic simulation. The method is different, but the results are essentially the same, and, given a long enough period, numerically the same. The frequency analysis approach is more time consuming but infinitely more informative. Note from the results that one can compare the frequency of an impact and the severity of the problem. This is useful in determining the risk of fish kills or the frequency of beach closures and solutions utilizing CSO controls to reduce such problems.

Performance Measures of CSO Controls

The Association of Metropolitan Sewerage Agencies (AMSA) with cooperative funding from the EPA developed a set of recommended performance measures

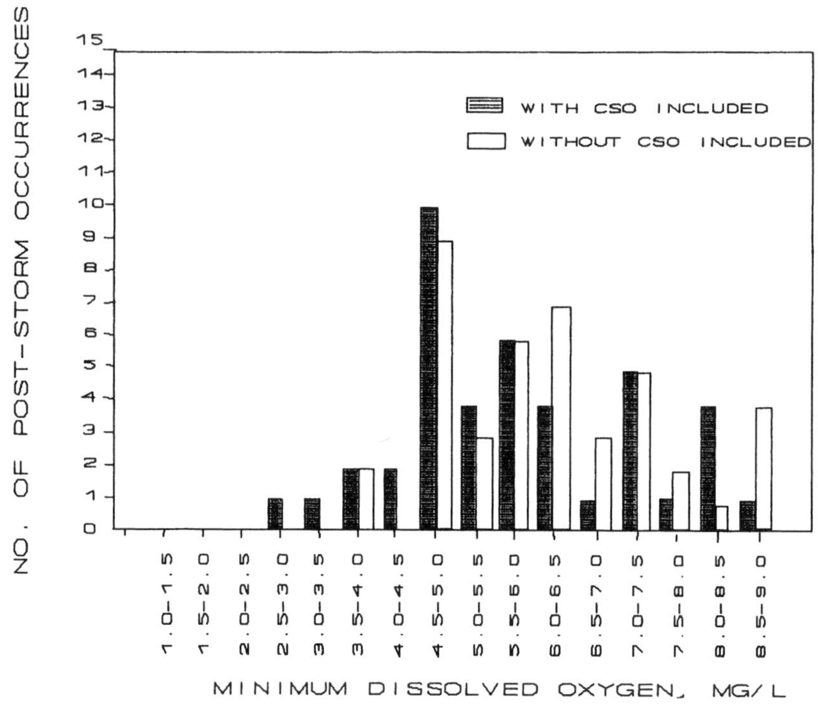

Figure 3-9. Predicted summer/postsummer dissolved oxygen in James River Estuary for a representative five-year period (Limno-Tech, Inc. 1987).

for tracking the success of CSO control efforts (AMSA January 1996). The AMSA performance measures include practical receiving-water, ecological, and human health benchmarking methods that may be selected, mixed, and matched to track progress for site-specific CSO situations and concerns. These performance measures are useful tools for assessing CSO control benefits and should be interactively selected with the water-quality impact evaluation methods to ensure consistency. In other words, the water-quality impact assessment techniques should be selected to enable municipalities to accurately forecast the benefits they will be tracking with their selected performance measures. The relevant AMSA CSO control performance measures are grouped as follows:

- **Receiving waters:** trends in dissolved oxygen, fecal coliform, floatables, sediment oxygen demand, and metals in bottom sediments.
- **Ecological/human health:** shellfish bed and beach closures; benthic organism and biological diversity indices.

Each of these water-quality, impact-related performance measures is discussed in this section with AMSA "profile sheets" provided. The profile sheets include a description, applicability and use, advantages, disadvantages, units, data sources, relative cost, and other important attributes.

Dissolved Oxygen Trend

As discussed previously, dissolved oxygen is vital to fish and other aquatic life and is needed to prevent the generation of odors. In receiving waters where dissolved oxygen levels are significantly impacted by CSO, time trends of measured concentrations may be very useful for identifying improvements caused by improved CSO controls that reduce the discharge of oxygen-demanding materials. Continuous automated measurements in known CSO impacted areas of the receiving waters provide the best record for tracking trends, but well-designed "grab" measurements may work as well. Figure 3–10 presents the AMSA profile sheet summarizing the attributes of this performance measure and provides an example.

Fecal Coliform Trend

Urban receiving waters typically have elevated levels of fecal coliform bacteria during and after storms due to discharges from a variety of sources, including CSOs in cities predominated by combined sewers. If elevated levels of fecal coliform are attributable to CSOs, then the reductions of these over time can be a significant measure of performance for implementation of CSO controls. Grab samples are adequate to characterize fecal coliform trends (in most receiving waters) if the samples are frequent and representative of *both* dry- and wet-weather conditions. The attributes and an example of this performance measure are provided in Figure 3–11.

Floatables Trend

Floatable materials in receiving waters are aesthetically displeasing and a potential public health problem for materials originating from CSOs. Reducing or eliminating this material can be an important and high-profile measure of progress in CSO controls that is understood by and meaningful to the public. The amount of floatable debris collected from waters and along the shore, if adequately quantified, can be tracked over time to observe trends that may be attributable to new CSO controls. However, floatable trend evaluations can be obscured or confounded because many of the same floatables in CSO are the same debris found in urban storm water. Also, collection of floatable debris is often done through volunteer efforts, where quantification is not precise. The AMSA performance measure profile sheet for floatable trends is presented in Figure 3–12.

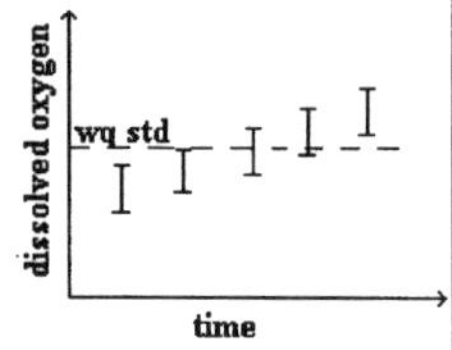

CSO Performance Measure Profile Sheet

Dissolved Oxygen Trend

Category: **Receiving Water**

Description:
Trend of dissolved oxygen in CSO impacted receiving water (time plot or trend analysis of dissolved oxygen in mg/l).

Applicability and Use:
The dissolved oxygen concentration of receiving waters is a widely used measure of water quality conditions. This receiving water measure has general applicability for receiving waters where it is consistent with the LTCP. It is most applicable in those situations where depressed dissolved oxygen conditions occur and are caused in part by CSO loadings and/or sediment oxygen demand originating from CSO solids.

Advantages of Measure:
- Dissolved oxygen can be monitored in the field with simple inexpensive equipment.
- Dissolved oxygen can be monitored continuously making it possible to record trends over time.
- Dissolved oxygen has value at the local and watershed levels.
- Historical records are available in many receiving waters.

Disadvantages of Measure:
- Impacts can be transferred downstream and away from CSO outfalls in some systems.
- It may be difficult to associate change over time with CSO reductions because of other watershed sources.

For More Information:
U.S. EPA. 1995. *CSOs: Guidance for Long-Term Control Plan.* EPA 832-B-95-002.

Units of Measurement:
Milligrams per liter (mg/l) of DO.

Probable Reporting Agency:
CSO Communities.
State and Federal Agencies.

Potential Data Sources:
- Water quality surveys.
- Continuous samplers.

Cost of Data Collection:
Medium to High. Requires long term commitment to consistent monitoring during wet weather events.

Attributes:

*Linkage to CSO discharge	◗
*Quantifiable for trend analysis	●
*Usefulness relative to the cost of data collection	◗
*Flexibility with respect to local objectives and conditions	◗
*Understandable to the public	◗
*Relationship to environmental improvement	●
*Watershed Context	●

KEY
● High ◗ Moderate ○ Low

Figure 3–10. AMSA profile sheet for dissolved oxygen trend performance measure.

Sediment Oxygen Demand Trend

Sediment oxygen demand (SOD) is a measure of the amount of dissolved oxygen in the receiving-water column that is removed and used by biologic decay of organic materials in the sediment. CSO may contain a high content of organic

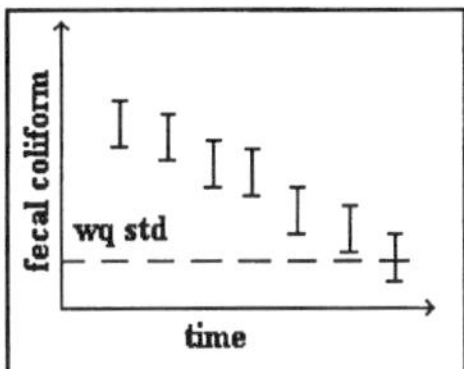

CSO Performance Measure Profile Sheet

Fecal Coliform Trend

Category: **Receiving Water**

Description:
Trend of the wet weather concentration of fecal coliform or other bacterial indicator concentrations in CSO-impacted receiving waters (time plot or trend analysis of fecal coliform in #/100 ml in receiving water).

Applicability and Use:
The level of fecal coliform in receiving waters is a widely used measure of water quality conditions. Bacterial contamination is one of the major pollutants of concern discharged from CSOs. High bacterial levels can affect shellfishing, swimming, and other designated uses. Measuring the trend of fecal coliform can be a useful way for many CSO communities to assess the effectiveness of CSO control measures.

Advantages of Measure:
- Widely applied to gauge attainment of designated uses and water quality standards.
- Historical records are available in many receiving waters.

Disadvantages of Measure:
- Cannot be continuously monitored, making it difficult to obtain wet weather samples during and following overflow events.
- It may be difficult to associate change over time with CSO control because of other watershed sources of contaminants and management activities.

For More Information:
APHA, AWWA, WEF. 1995. *Standard Methods For the Examination of Water and Wastewater, 19th Edition.*

Units of Measurement:
per 100 ml or other appropriate units.

Probable Reporting Agency:
CSO Communities.
State Agencies.

Potential Data Sources:
- Water quality surveys.

Cost of Data Collection:
Medium to High. Requires long term commitment to consistent monitoring during wet weather events.

Attributes:

* Linkage to CSO discharge	◗
* Quantifiable for trend analysis	●
* Usefulness relative to the cost of data collection	◗
* Flexibility with respect to local objectives and conditions	◗
* Understandable to the public	◗
* Relationship to environmental improvement	●
* Watershed Context	●

KEY
● High ◗ Moderate O Low

Figure 3-11. AMSA profile sheet for fecal coliform trend performance measure.

materials that settle to the bottom and thereby contribute to SOD. Sediment oxygen demand is important mainly in sluggish systems where low dissolved-oxygen conditions occur. Specialized measurements of SOD can be conducted and changes tracked over time for receiving waters impacted by CSOs. Consistency in monitoring locations and measurement techniques is essential for a

CSO Performance Measure Profile Sheet

Floatables Trend

Category: Receiving Water

Description:
Trend of floatables measured or collected from receiving water cleanup programs or beach surveys (cubic yards, pounds or tons per year).

Applicability and Use:
Floatables are one of the major pollutants of concern discharged from CSOs. CSO communities are expected to control floatables as one of the Nine Minimum Controls. Tracking the trend of floatables collected in key use areas can be a useful performance measure for some CSO communities where it is consistent with the LTCP. Floatables trend represents an alternative to the end-of-pipe capture and measurement of floatables.

Advantages of Measure:
- Understandable to the public.
- Can be targeted at key problem areas to reduce cost and effort.
- Can help to determine the CSO contribution to debris problems.

Disadvantages of Measure:
- May be difficult to detect trend as volume of floatables may be highly variable.
- Non-CSO sources may still be a problem following CSO control.

For More Information:
U.S. EPA. 1995. *CSOs: Guidance for Nine Minimum Controls.* EPA 832-B-95-003.

Units of Measurement:
Cubic yards, pounds, tons per year or number of items collected.

Probable Reporting Agency:
CSO Communities.
Environmental Agencies.
Volunteers.

Potential Data Sources:
- Monthy or annual survey.
- Cleanup program records.

Cost of Data Collection:
Variable. Use of volunteers is low. Use of equipment intensive methods such as skimmer boats is high.

Attributes:

*Linkage to CSO discharge	◗
*Quantifiable for trend analysis	◗
*Usefulness relative to the cost of data collection	◗
*Flexibility with respect to local objectives and conditions	●
*Understandable to the public	●
*Relationship to environmental improvement	◗
*Watershed Context	◗

KEY
● High ◗ Moderate O Low

Figure 3–12. AMSA profile sheet for floatables trend performance measure.

meaningful trend analysis for SOD and for relating observed changes to CSO controls. This information has been useful in identifying improvements in some systems as seen in Figure 3–13 below for the Chicago River (AMSA January 1996) and in the AMSA profile sheet in Figure 3–14.

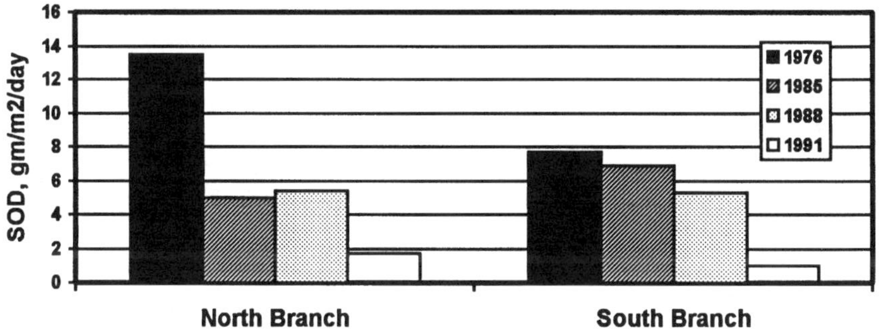

Figure 3–13. Trends in SOD at CSO impacted locations in the Chicago River (AMSA January 1996).

Metals in the Sediments Trend

Trace metals, when present in CSO discharges, are often adsorbed to particles or are in particulate form. When discharged to sluggish receiving-water systems, these particulate fractions settle and may accumulate to potentially harmful levels over time. The metals most commonly found in urban runoff, and hence in CSO, are cadmium, chromium, copper, lead, nickel, and zinc. When excess levels of sediment metals have been identified in the vicinity of CSO outfalls, then tracking these levels annually may be useful as a performance measure for identifying the benefits of newly implemented CSO controls. Specialized skills are needed for conducting these measurements, and monitoring consistency from year to year is essential for identifying trends as indicated in the profile sheet (Figure 3–15).

Shellfish Bed and Beach Closures

Shellfishing and contact recreation such as swimming and wading are major recreational uses for many CSO receiving waters. When beaches and shellfish beds lie near CSO outfalls, closings due to bacteriological and toxics concerns may be attributable to CSO pollutants, depending on local currents and dilution. Care is warranted in trying to draw cause-and-effect relationships between CSO discharges and closings because other pollution sources can play a major role and may even offset improvements gained by CSO controls. However, where closings are clearly attributable to CSO discharges, reductions in closings can be an important goal and measure of performance for CSO control programs. Improvements have direct recreational benefits for local citizens. Examples of these performance-measure applications are in Figures 3–16 and 3–17—the AMSA CSO performance-profile sheets for beach and shellfish bed closings.

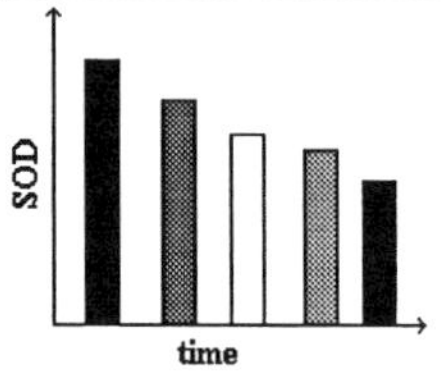

CSO Performance Measure Profile Sheet

Sediment Oxygen Demand Trend

Category: Receiving Water

Description:
Trend of sediment oxygen demand in CSO-impacted receiving waters (time plot of sediment oxygen demand).

Applicability and Use:
Sediment Oxygen Demand (SOD) is a measure of the capacity of bottom sediments to remove oxygen from overlying waters. The performance measure may be useful in receiving waters where dissolved oxygen levels are continuously depressed, and where CSOs have been identified as contributing substantially to organic loadings. Tracking SOD over time may be an effective way to quantify long-term improvements due in part to CSO control.

Advantages of Measure:
- Trend can be linked to CSO control in some CSO communities.
- Measurement once per year should be sufficient.

Disadvantages of Measure:
- Requires high skill level to measure Sediment Oxygen Demand properly.
- Sediment Oxygen Demand is not widely understood.
- Sediment Oxygen Demand is not necessarily meaningful in itself. Its importance varies with its contribution to the overall dissolved oxygen budget of receiving water.

For More Information:
APHA, AWWA, WEF. 1995. *Standard Methods For the Examination of Water and Wastewater, 19th Edition.*

Units of Measurement:
Grams of O_2 in meters squared/day.

Probable Reporting Agency:
State Agencies.
Research institutions.

Potential Data Sources:
- Annual survey.

Cost of Data Collection:
Medium to High. Measurement requires specialized field and laboratory expertise.

Attributes:

* Linkage to CSO discharge	◗
* Quantifiable for trend analysis	●
* Usefulness relative to the cost of data collection	◗
* Flexibility with respect to local objectives and conditions	◗
* Understandable to the public	○
* Relationship to environmental improvement	●
* Watershed Context	●

KEY
● High ◗ Moderate ○ Low

Figure 3–14. AMSA profile sheet for sediment oxygen demand trend performance measure.

Benthic Organism Index

Benthic (sediment) organisms in receiving waters are an important food source for fish and respond over time to the integrated effects of pollutant stresses from all sources. The available indices characterizing the abundance

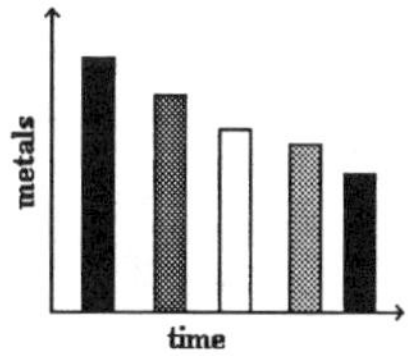

CSO Performance Measure Profile Sheet

Trend of Metals in Bottom Sediments

Category: Receiving Water

Description:
Trend of the concentration of selected metals in bottom sediments (time plot of metals in mg/kg).

Applicability and Use:
Elevated metals concentration in bottom sediments may cause human health and ecological impacts. This performance measure may be useful in areas where sediment contamination is present, and where the reduction of metals load from the CSS is a goal of the LTCP.

Advantages of Measure:
- Trend cam be linked to CSO control in some CSO communities.
- Measurement once per year should be sufficient.

Disadvantages of Measure:
- Requires a high skill level to measure metals in bottom sediment properly.
- Non-CSO sources of metals contamination are typically dominant in most areas.
- It may be difficult to associate change over time with CSO control because of other watershed sources of pollution.

For More Information:
APHA, AWWA, WEF. 1995. *Standard Methods For the Examination of Water and Wastewater, 19th Edition.*

Units of Measurement:
Metals concentration in mg/kg.

Probable Reporting Agency:
State Agencies.
CSO Communities.

Potential Data Sources:
- Annual survey.

Cost of Data Collection:
Medium to High. Measurement requires specialized field and laboratory expertise.

Attributes:

* Linkage to CSO discharge	○
* Quantifiable for trend analysis	●
* Usefulness relative to the cost of data collection	◗
* Flexibility with respect to local objectives and conditions	◗
* Understandable to the public	○
* Relationship to environmental improvement	●
* Watershed Context	●

KEY
● High ◗ Moderate ○ Low

Figure 3-15. AMSA profile sheet for metals in the sediments trend performance measure.

and diversity of organisms in the benthos are therefore excellent measures of the ecological health of receiving waters. If CSO is a major component of the pollutant discharges to a stressed water body, then tracking annual changes in the benthic organisms by aquatic biologists can be a fundamental measure of the performance of CSO controls implementation to improve water quality.

CSO Performance Measure Profile Sheet

Beach Closures

Category: Ecological/Human Health/ Resource Use

Description:
Number of beach closures attributed to CSO discharges (# per year).

Applicability and Use:
Beach closures represent a serious impairment of a designated use. Beach closures are usually based upon high bacterial levels or the presence of potential health hazards due to the wash-up of floatables. Both bacterial contamination and floatables are major pollutants of concern discharged from CSOs. This performance measure may be appropriate for CSO communities where beach closures are a problem, and where CSO sources contribute to the problem.

Advantages of Measure:
- Widely used and understood by the public.
- Directly linked to the "swimmable" goal of the Clean Water Act.

Disadvantages of Measure:
- It may be difficult to associate change over time with CSO control because of other watershed sources of pollution and management activities.

For More Information:
National Resources Defense Council. 1995. *Testing the Waters V: Politics and Pollution at U.S. Beaches.*

Units of Measurement:
Number per year.

Probable Reporting Agency:
State/County Health Department.

Potential Data Sources:
- Local records.

Cost of Data Collection:
Low to Medium. Data should be readily available from Health Departments.

Attributes:

* Linkage to CSO discharge	◗
* Quantifiable for trend analysis	●
* Usefulness relative to the cost of data collection	●
* Flexibility with respect to local objectives and conditions	◗
* Understandable to the public	●
* Relationship to environmental improvement	●
* Watershed Context	●

KEY
● High ◗ Moderate ○ Low

Figure 3–16. AMSA profile sheet for beach closures performance measure.

Again, care must be taken to avoid attributing non-CSO benthic impacts to CSOs. This can be accomplished by sampling locations inside and outside the obvious influence of the CSOs but similar in all other respects. More details about the benthic indices are provided in the AMSA profile sheet (Figure 3–18).

CSO Performance Measure Profile Sheet

Shellfish Bed Closures

Category: Ecological/Human Health/ Resource Use

Description:
Extent of shellfish bed closures attributed to CSO discharges (acre/days closed per year).

Applicability and Use:
Shellfish bed closures pose a serious impact to commercial and recreational activities. Shellfish bed closures are based upon bacterial levels. Bacterial contamination is one of the major pollutants of concern discharged from CSOs. This performance measure may be appropriate for CSO communities where shellfish bed closures are a problem, and where CSO sources contribute to the problem.

Advantages of Measure:
- Widely used and understood by the public.
- Represents a direct measure of a designated use.
- Historical records are available in many receiving waters.

Disadvantages of Measure:
- It may be difficult to associate change over time with CSO control because of other watershed sources of pollution and management activities.

For More Information:
U.S. Food and Drug Administration (FDA). 1992. *National Shellfish Sanitation Program Manual of Operations, Part I: Sanitation of Shellfish Growing Areas.*

Units of Measurement:
Acre/days closed per year.

Probable Reporting Agency:
State/County Health Department.

Potential Data Sources:
- Health Department Reports.

Cost of Data Collection:
Low to Medium. Data should be available from Health Departments.

Attributes:

* Linkage to CSO discharge	◗
* Quantifiable for trend analysis	●
* Usefulness relative to the cost of data collection	●
* Flexibility with respect to local objectives and conditions	◗
* Understandable to the public	●
* Relationship to environmental improvement	●
* Watershed Context	●

KEY
● High ◗ Moderate O Low

Figure 3-17. AMSA profile sheet for shellfish bed closures performance measure.

Biological Diversity Index

A more general measure of ecosystem health for CSO receiving waters is provided by applying a biological diversity index such as the Index of Biological Integrity (IBI), Index of Well Being (I_{wb}), or the EPA's Rapid Bioassessment Protocol (RBP). These indices provide indexed measures of the abundance and

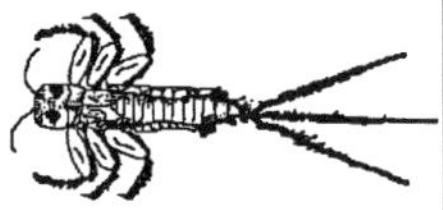

CSO Performance Measure Profile Sheet

Benthic Organism Index

Category: **Ecological/Human Health/ Resource Use**

Description:
Annual survey of macroinvertebrates in CSO receiving waters (e.g., Rapid Bioassessment Protocols (RBP), Macroinvertebrates Biotic Index (MBI), or Invertebrate Community Index (ICI)).

Applicability and Use:
The use of benthic organism indices as a measure of the health of aquatic systems in rising. Biological measures are being integrated with and in some instances replacing traditional water quality measures. This performance measure may be applicable in impaired water where CSO control and other environmental management activities are expected to improve aquatic habitat over time.

Advantages of Measure:

- Provides a direct measure of ecological results.
- Measurement once per year should be sufficient, if performed during the same season each year. Longer intervals may be acceptable.

Disadvantages of Measure:

- Requires high skill level to measure properly.
- Results are not readily understandable to the public, or even to environmental professionals other than specialized biologists.

For More Information:
U.S. EPA. 1989. *Rapid Bioassessment Protocols (RBP) for Use in Streams and Rivers - Benthic Macroinvertebrates and Fish.* EPA 440-4-89-001.

Units of Measurement:
Method specific index.

Probable Reporting Agency:
State Agencies.

Potential Data Sources:

- Annual surveys.

Cost of Data Collection:
Medium to High. Measurement requires specialized field and laboratory expertise.

Attributes:

* Linkage to CSO discharge	◗
* Quantifiable for trend analysis	●
* Usefulness relative to the cost of data collection	●
* Flexibility with respect to local objectives and conditions	●
* Understandable to the public	○
* Relationship to environmental improvement	●
* Watershed Context	●

KEY
● High ◗ Moderate ○ Low

Figure 3–18. AMSA profile sheet for benthic organism index performance measure.

diversity of multiple aquatic organisms, including those of the benthos. Since they are applied on a widespread basis, they are more prone to confounding influences from other non-CSO pollutant sources but can be very useful for watershed-wide evaluations. Applied with care to waters obviously impaired by CSOs, these indices can be a useful measure of the performance of CSO controls. The details for applying this measure to CSO are provided in Figure 3–19.

CSO Performance Measure Profile Sheet

Biological Diversity Index

Category: Ecological/Human Health/ Resource Use

Description:
Annual survey of aquatic communities in CSO receiving waters (e.g., Index of Biologic Integrity (IBI), Rapid Bioassessment Protocol (RBP) etc.).

Units of Measurement:
Method specific index.

Probable Reporting Agency:
State Agencies.

Applicability and Use:
The use of biological diversity indices as a measure of the health of aquatic systems is rising. Biological measures are being integrated with and in some instances replacing traditional water quality measures. This performance measure may be applicable in impaired water where CSO control and other environmental management activities are expected to improve aquatic habitat over time.

Potential Data Sources:
- Annual surveys.

Cost of Data Collection:
Medium to High. Measurement requires specialized field and laboratory expertise.

Advantages of Measure:
- Provides a direct measure of ecological results.
- Measurement once per year should be sufficient, if performed during the same season each year.
- Directly related to the "fishable" goal of the CWA.

Disadvantages of Measure:
- Requires high skill level to measure properly.
- Results are not very understandable.
- Drought can affect results in some water bodies.
- Results are not readily understandable to the public, or even to environmental professionals other than specialized biologists.

For More Information:
U.S. EPA. 1989. *Rapid Bioassessment Protocols (RBP) for Use in Streams and Rivers - Benthic Macroinvertebrates and Fish.* EPA 440-4-89-001.

Attributes:

* Linkage to CSO discharge	◗
* Quantifiable for trend analysis	●
* Usefulness relative to the cost of data collection	●
* Flexibility with respect to local objectives and conditions	●
* Understandable to the public	○
* Relationship to environmental improvement	●
* Watershed Context	●

KEY
● High ◗ Moderate ○ Low

Figure 3–19. AMSA profile sheet for biological diversity index performance measure.

Case Studies

The previous discussions have presented generalized considerations and analyses used for CSO impact assessments. However, experience and specific examples are often the best teachers. Therefore, four short discussions follow that highlight case histories involving water-quality evaluations for CSO control. The focus of these discussions is on the methods, not the specific results or the recommended CSO controls. The reader is encouraged to take note of the differences in methodologies as they relate to individual characteristics of the study site.

Portland, Oregon

The Willamette River drains 60% of the State of Oregon's drainage basin and encompasses nearly 80% of the state's population. The sewer system in Portland is designed to convey up to three times the base sanitary flow and captures and treats approximately 96% of the sewage from homes and businesses. The other 4 percent becomes part of the untreated overflow discharged at 42 outfalls on the Willamette River and 13 outfalls on the Columbia Slough. During a typical year, there are approximately 150 days of rainfall in Portland. The magnitude and frequency of overflow varies from one outfall to another. Some outfalls overflow virtually every time it rains, whereas others overflow as few as 30 days in a typical year. During an average year, Portland's combined sewer system discharges an estimated total of six billion gallons of urban storm water mixed with sewage. The major near-field CSO water-quality concern in the Willamette River is the possibility of acute impacts of trace metals during and immediately after a storm.

One approach for assessing Portland CSO impacts on receiving-water quality was to simulate the near-field impacts of CSO discharges on receiving-water quality. The U.S. EPA mixing zone model (PDS) was calibrated and applied to individual CSO outfalls for calculating near-field dilution. PDS is a three-dimensional, steady-state single discharge model that predicts the dilution, assuming relatively constant localized river and discharge conditions. The model calculates the minimum dilution along the centerline, the average lateral dilution, a plume width, and a plume depth. This model was used to assess the potential frequency and extent of near-field metal contamination in the water column due to CSO discharges.

The PDS mixing zone model application required inputs describing the outfall configuration and the ambient river conditions. The width, water depth, overflow flow velocity, water temperature, and outfall orientation (angle with the general river flow direction) are the major outfall configuration parameters. Model ambient stream parameters included salinity, heat conduction coefficient, horizontal and vertical dispersion coefficients, jet entrainment, and water temperature.

Validation of the PDS model was conducted at two individual CSO outfalls to verify its ability to accurately forecast near-field dilution of CSO discharges. Validation tests were conducted by injecting tracer dye into a manhole downstream of the overflow weir and then tracking the dye in the receiving waters along horizontal transects and vertical profiles in the vicinity of the outfall. Both model predictions fit extremely well with the observed data.

Following validation, the PDS mixing zone model was used to forecast the water-quality conditions in the acute mixing zones of all individual CSO outfalls. Model simulations provided minimum dilution estimates along the plume centerline, areas at or below various dilution values, distance downstream of outfalls to reach required dilution, and the distance to complete mix. Figure 3–20 presents the model output results for an individual outfall, No. 38, showing the percentage of effluent in the plume, given the surface area within the plume and the distance along the plume centerline.

The model simulations for the predicted surface area of acute-criteria exceedances for three trace metals due to CSO discharges were totaled for all individual outfalls. Figure 3–21 presents the total surface area of acute-criteria exceedance as a percentage of the total river surface area within the CSO area (2,700 acres). The predicted relative areal impact of CSO discharges was very small for all three metals (less than 1%) when compared to the total area of the river.

South Bend, Indiana

South Bend's combined sewer system serves 14,000 acres and over 90% of the city's 109,000 people. Two cities upstream of South Bend also have CSOs. Mishawaka is just upstream of South Bend, and Elkhart is approximately 12 river miles upstream of South Bend. CSOs discharge to the St. Joseph River, which is very important to the economic development of South Bend and an important recreational resource. A comprehensive study was performed to develop a cost-effective CSO control plan that effectively balances control technologies, costs, river-water quality, and river uses.

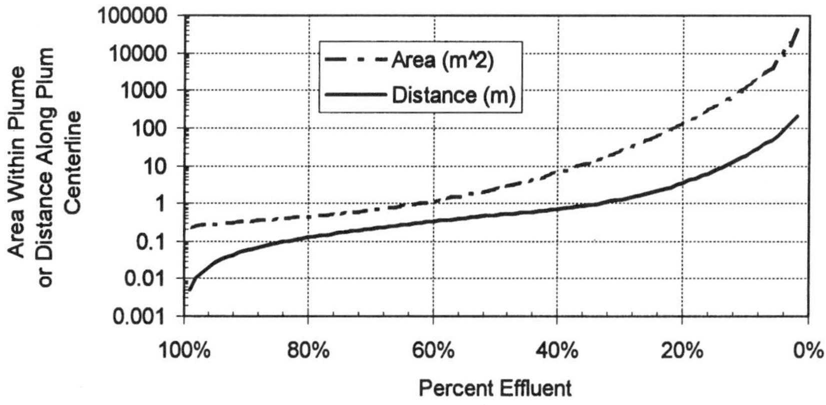

Figure 3–20. Modified PDS Mixing Zone Modeling Results for Portland, Oregon, CSO outfall No. 38.

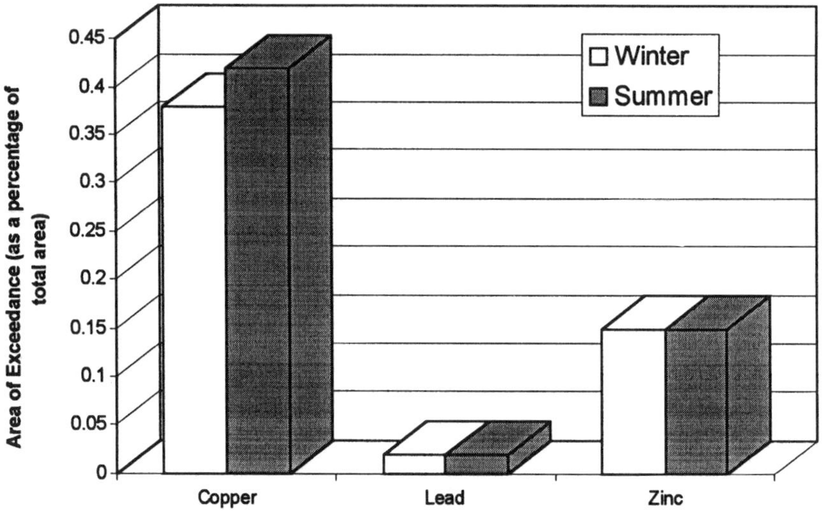

Figure 3-21. Mixing zone modeling results for Portland, Oregon, CSOs.

The South Bend CSO study (as described in Dillon et al. 1994) was conducted in anticipation of the final national CSO control policy. It included monitoring and modeling of CSO impacts and control alternatives to evaluate and select a cost-effective control plan.

The river computer model PULSEQUAL was developed and calibrated to estimate the dynamic effects of CSOs on dissolved oxygen and bacteria level in the river. PULSEQUAL inputs and coefficients were calibrated to the St. Joseph River using wet-weather data from the summer of 1991. Continuous simulations were run to forecast river impacts in response to 10 years of representative rainfall. The 10-year CSO pollutant loads were estimated based on forecasts from a calibrated sewer system model.

Three initial CSO control scenarios were evaluated. The first scenario was existing sewer and drainage-basin conditions. The second scenario assumed that all CSOs in South Bend were eliminated. The third scenario examined the effect if all CSOs to the river were eliminated including those upstream. The second scenario is especially important because it determined the incremental contributions of the CSOs and sources outside South Bend to the existing water-quality conditions. The third scenario defined the theoretical maximum water-quality benefits for eliminating all CSOs in and upstream of South Bend. Water-quality forecasts were evaluated for the critical locations in the river at the point of greatest impacts. (Forecasted bacteria and dissolved oxygen impacts were less at all other locations in the river.)

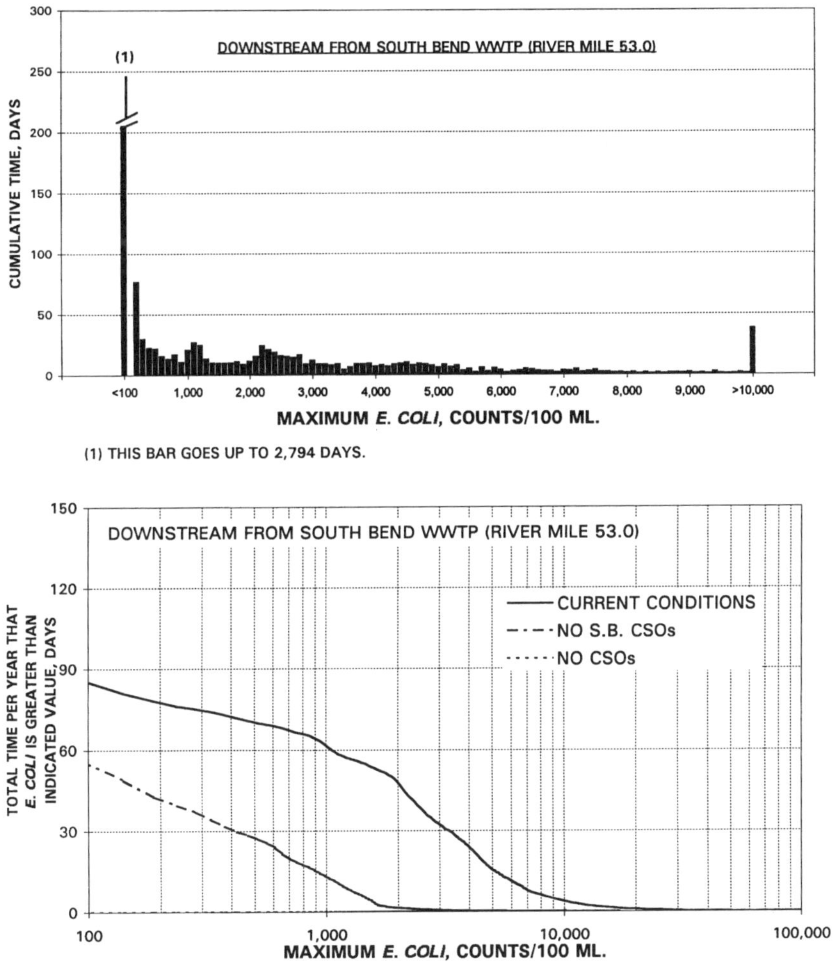

Figure 3–22. PULSEQUAL forecasts of *E. coli* impacts on the St. Joseph River from South Bend, Indiana, and other CSOs.

Figure 3–22 shows the results of the 10-year PULSEQUAL river forecasts for *E. coli* bacteria at the critical location (river mile 53). The upper panel of the figure shows the forecasted cumulative time during the 10 years *E. coli* concentrations will be exceeded at the particular concentration levels under existing conditions. *E. coli* concentrations were forecasted to remain below 100 counts per 100 ml for 2,794 days (76.5 percent of the time), and the remaining 856 days were distributed at higher concentrations ranging up to over 10,000 counts per 100 ml. This location was predicted to experience a cumulative total of 760 days in 10 years (76 days average per year) when the *E. coli* levels are higher than the *E. coli* water-criteria maximum of 235 counts per 100 ml. This corresponds to a total of 1,320 individual occurrences over the maximum criteria during the 10 years (132 times per year on average). Therefore, the average duration of an excess of *E. coli* exceedance is approximately 13.8 hours for each occurrence at the critical location.

Figure 3–22, lower panel, presents the current conditions in an annual average format and compares them to results for complete elimination of South Bend CSOs. Elimination of South Bend CSOs ("No S.B. CSOs") produces a significant reduction in the total time per year the elevated *E. coli* levels occur throughout the range of simulated concentrations. However, substantially elevated concentrations still occur due to upstream CSO sources. For example, under current conditions, bacterial concentrations in excess of 1,000 counts per 100 ml are forecasted to occur approximately 1,440 hours (60 days) per year on average but would occur only about 288 hours (12 days) per year without the influence of South Bend CSOs. South Bend CSOs are a major source of *E. coli* bacteria problems but not the only source of *E. coli.* Elimination of all the CSOs in South Bend and the CSOs upstream of South Bend ("No CSOs") is represented by the bottom line that indicated there will be no significant *E. coli* concentrations under this scenario.

In contrast to the bacteria forecasts, the 10-year dissolved oxygen simulations indicated that deoxygenation due to CSO pollutants is nominal in the St. Joseph River. As shown in Figure 3–23, the forecasted minimum dissolved oxygen at the critical location in the river is nearly always above the dissolved oxygen criteria minimum of 6 mg/l under existing conditions. The annual average dissolved-oxygen forecasts with and without all CSOs (including South Bend) are nearly indistinguishable at the critical location in the river. It is clear that CSO controls offer very minor potential dissolved-oxygen benefits (the narrow area between the forecasted "No CSOs" and "Current Conditions" lines on Figure 3–23) and were not a major factor in the evaluation and selection of control alternatives.

The river model forecasts indicated that controlling South Bend CSOs would provide the greatest benefits in reducing *E. coli* concentrations but would not appreciably affect dissolved-oxygen levels in the St. Joseph River.

A cost-performance analysis was performed to establish the level of CSO control (design storm) and to select the most cost-efficient, long-term CSO control plan. The river model was used to estimate the impacts of CSOs on river *E. coli* for the 10-year rainfall period for different control levels. Control levels included no control and the capture, storage, and treatment of the storms shown in the table at the bottom of the following page.

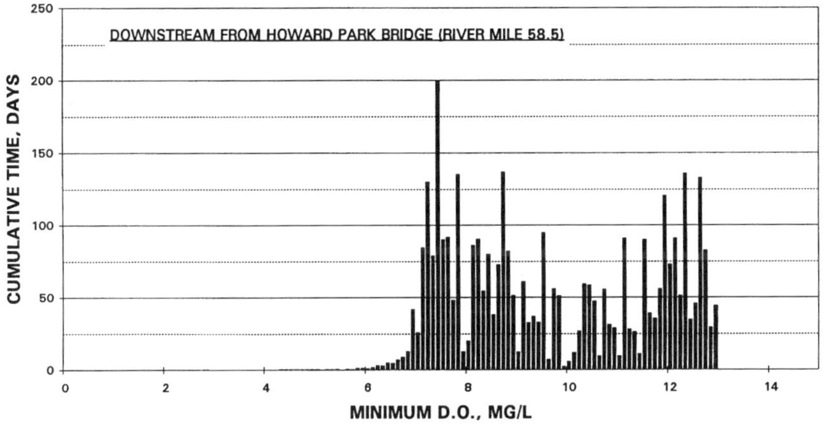

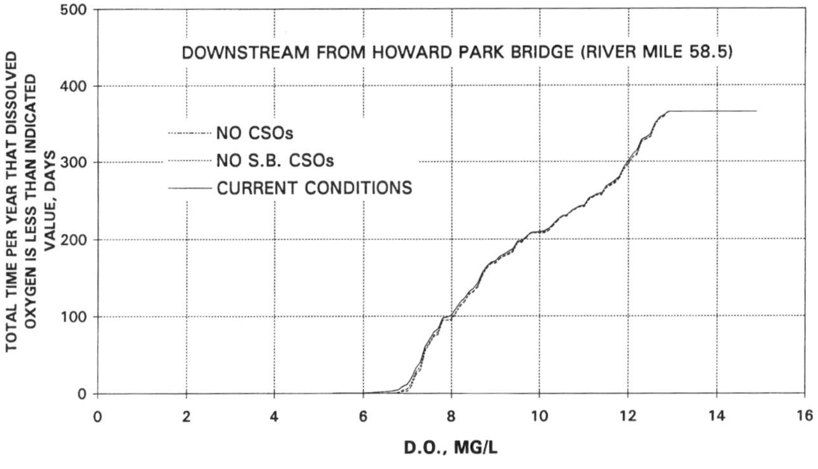

Figure 3-23. PULSEQUAL forecasts of dissolved-oxygen impacts on the St. Joseph River from South Bend, Indiana, and other CSOs.

Storm	Total Rainfall (Inches)	Percent of Storms with Total Rainfall Equal to or Less than
C	0.24	65
D	0.49	80
E	0.80	90
F	1.23	95
G	2.00	99

Indiana surface waters must meet certain minimum water-quality criteria. Current state bacterial standards are oriented to dry-weather conditions and do not address wet weather. Also, the state standards were established for certain river uses that are not possible during wet weather. The fast-moving St. Joseph River precludes swimming. Therefore, an *E. coli* criterion of 576 counts per 100 ml was used reflecting "infrequently used full body contact" recreation (EPA 1986).

Figure 3–24 shows the cost-performance curve, estimated total time of exceedance per year of the selected *E. coli* criterion for the different levels of CSO control versus the capital costs for the different levels of control. (It was assumed that the CSOs upstream of South Bend receive the same level of CSO control.) On the cost-performance curve, the most cost-effective system occurs where the increment of benefit achieved in the river diminishes compared to the increased capital cost. This is referred to as the "knee of the curve." The "knee of the curve" for the St. Joseph River was determined to be between Storm C and Storm D at a total rainfall of approximately 0.40 inch. Capturing and treating a 0.40-inch storm reduces the forecasted total time of exceedance of *E. coli* concentration of 576 counts/100 ml from an estimated 70 days to 4.9 days (a 93% reduction). This curve was used as a tool to select affordable and effective controls for South Bend's LTCP.

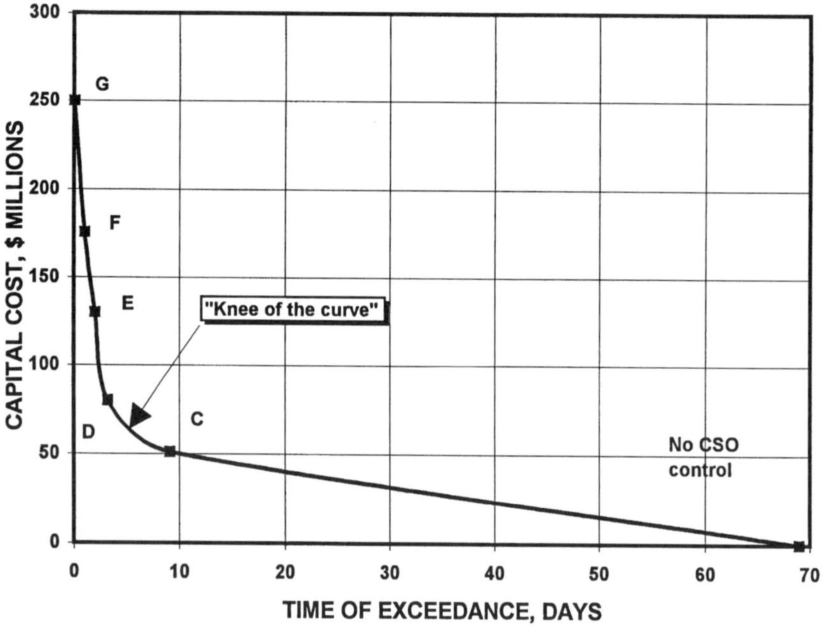

Figure 3–24. Cost-performance curve for South Bend CSO controls to meet "infrequently used full body contact recreation" criteria in the St. Joseph River.

Washington, DC

From 1978 to 1983, the District of Columbia studied its combined sewer system to develop control alternatives that were both environmentally protective and cost effective (O'Brien and Gere 1982). Since then, some programs for CSO control were implemented with varying success. The District's combined sewer system includes over 125,000 acres of sewer network with land use almost entirely urbanized or densely residential. For 50–75 storms per year, overflows discharged to three very different bodies of water (Figure 3–25). The largest quantity of overflows (approximately 3,200 MG/yr.) were discharged into the Anacostia River, a tidal-dominated tributary to the Potomac Estuary. The mean advective flow in the Anacostia is 130 cfs. Rock Creek received the smallest quantity of overflows (125 MG/yr.) but has a smaller mean flow (55 cfs); it is primarily surrounded by park areas. The upper Potomac Estuary received the remainder of the discharges directly (1,700 MG/yr.), plus rapid flushing of CSOs

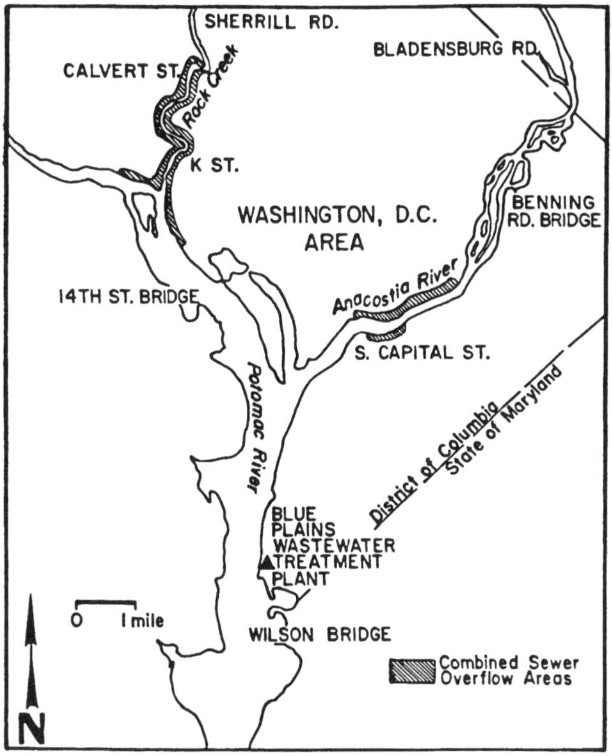

Figure 3–25. Washington, D.C., Study Area (Limno-Tech, Inc. 1982).

from Rock Creek, tidal exchange from the Anacostia, and wastewater-treatment-plant discharges and bypasses from the Blue Plains Wastewater Treatment Plant.

A two-year intensive CSO and receiving-water-quality sampling program was conducted to document problems and develop models for analyzing the effectiveness of different control alternatives (Limno-Tech, Inc. September 1982). Six continuous-recording, dissolved-oxygen monitors were used to document the frequency and severity of wet-weather impacts. Two intensive water-quality surveys were conducted on each receiving stream prior, during, and following wet-weather events. Sampling in the estuaries ranged from three- to five-hour intervals for five days, whereas in Rock Creek, samplings were every two to four hours for two days because of the shorter time of passage. Dry-weather surveys were conducted to document baseline conditions. Sampling focused on BOD, nitrogen, dissolved oxygen, solids, and bacteria, but less intensive sampling was also conducted for other parameters such as phosphorus, trace metals, and sediments.

The modeling approaches used for the three receiving streams were each different, reflecting their different characteristics (Limno-Tech, Inc. September 1982). Each model was calibrated and verified to survey data. The Dynamic Estuary Model (DEM) was used to simulate Potomac and Anacostia river conditions. This complex model has both hydrodynamic and water-quality simulation capabilities that were necessary to describe the tidal mixing in the two estuaries. For Rock Creek, the AUTO-QD stream model was used. This model allowed for dynamic water-quality calculations but did not provide hydrodynamic calculations. The absence of tides and the rapid movement of water in Rock Creek permitted a reasonable approximation of flows and velocities using a mass balance approach. This allowed for quicker and easier computations. Both models were used in a continuous simulation mode to simulate ten years of wet-weather, water-quality impacts.

Impacts in the three water bodies were very different. In the Potomac Estuary, dissolved-oxygen impacts were typically very small (less than 0.5 mg/l) because of large dilution and tidal mixing. However, CSO fecal coliform contributions caused approximately 40 extra days per year exceeding contact recreational criteria. The Anacostia River had similar impacts on bacteria but at much higher concentration levels. However, in contrast to the Potomac, dissolved oxygen was severely impacted in the Anacostia due to lower flow and tides. Prior to controls, 30–40 days per year were predicted to have oxygen concentrations below 1.0 mg/l due to CSO and sediment demand (caused largely by CSO). For Rock Creek, the situation was even more different. Dissolved-oxygen impacts were insignificant because of high reaeration and fast flushing. Bacterial impacts were frequent (approximately 50 days/year) but very short lived.

Data and models were used in this study to compare different CSO control strategies based on the cost and benefits expressed in terms of reduction of violations in water-quality and recreational standards. Shown in Figure 3–26 is an ex-

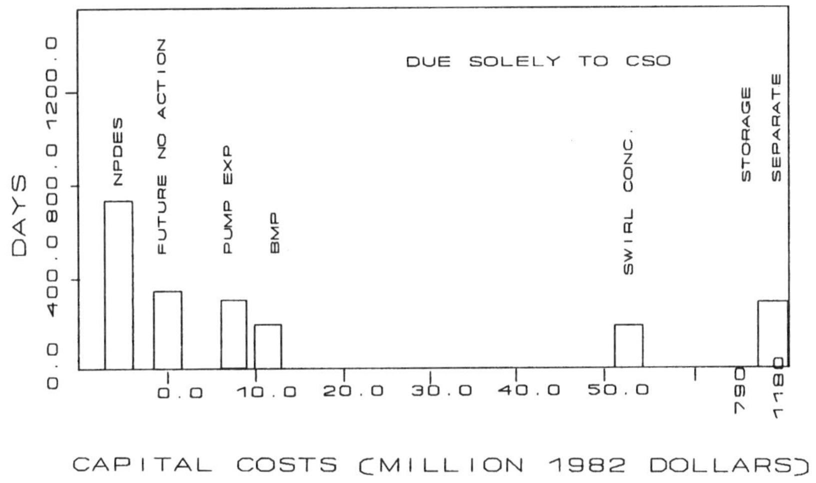

Figure 3–26. Comparison of district CSO control options to reduce Potomac Estuary fecal coliform (*E. coli*) days in 10 years with *E. coli* above 20 no./100 ml (Limno-Tech, Inc. September 1982).

ample of how different CSO control alternatives were compared in terms of how they reduced the frequency of water-quality problems (fecal coliform bacteria). Based on these and other study results, the District of Columbia selected a CSO control program that is now mostly implemented. Water-quality improvements and CSO reductions have been observed.

Poughkeepsie, New York

The city of Poughkeepsie is located 76 miles upstream on the Hudson River from New York City (Figure 3–27). It has 562 acres served by combined sewers, plus 1,814 acres served by separate sewers. However, both combined- and separate-sewered areas are served by a common interceptor. Overflows from individual areas, as well as pump-station bypasses, fed into the Hudson River in a strongly tidal area where stage variations are approximately four feet. Overflows typically averaged 1–2 MG but sometimes exceeded 5 MG. Monitoring and modeling studies were used to compare 12 different CSO control options (Limno-Tech, Inc. 1983; O'Brien and Gere 1983).

Two wet-weather and two dry-weather water-quality surveys were conducted to document water-quality problems. Six locations were sampled for BOD, Kjeldahl nitrogen, solids, dissolved oxygen, fecal coliform bacteria, and several other secondary parameters. Sewer-overflow and wastewater-effluent monitoring was conducted in conjunction with the water-quality surveys. Wet-

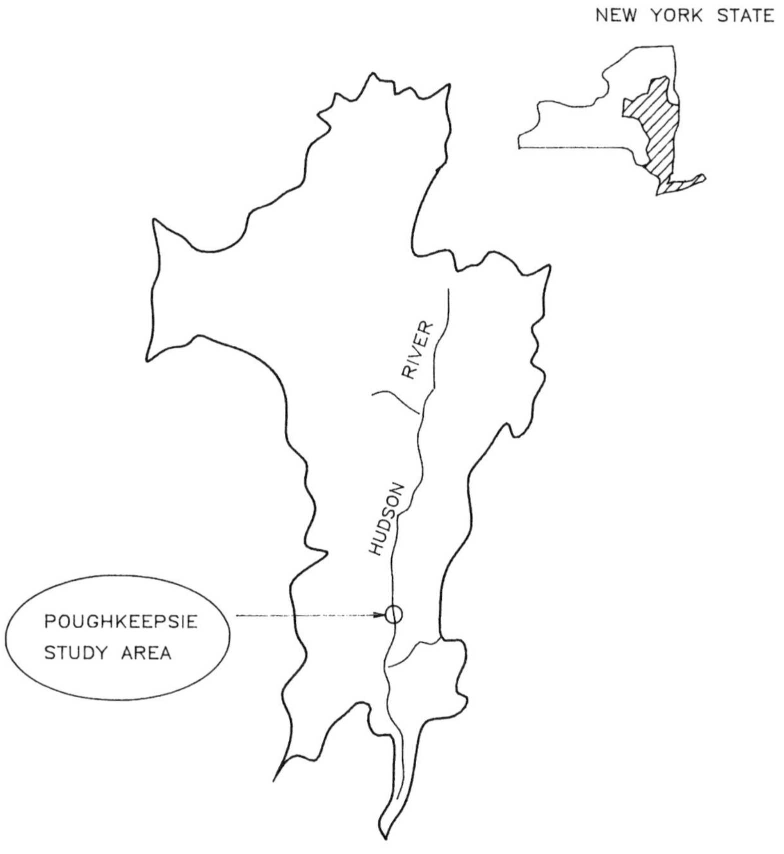

Figure 3–27. Location of Poughkeepsie, New York (Limno-Tech, Inc., August 1983).

weather impacts on the Hudson River were simulated using the analytical model described earlier in the chapter (on estuaries). This model simulated estuary response due to a "pulse" loading. It accounted for dilution, tidal dispersion, and reaction. The proximity of overflows and short duration allowed for the use of this easy computational method. The model was linked with a Monte Carlo simulation routine to calculate the frequency of water-quality problems based on probability distributions for river flow, temperature, and CSO loadings. Shown in Figure 3–28 is the summary of predicted fecal coliform concentrations for all storms simulated. Monte Carlo forecasts were run for each of 12 control alternatives, and costs were compared to benefits in terms of reduced violations.

Monitoring and model results showed moderate impacts of Poughkeepsie CSO on Hudson River fecal-coliform levels. Concentrations were predicted to exceed 1,000 counts/100 ml for 7–15 days per year and 200 counts/100 ml for

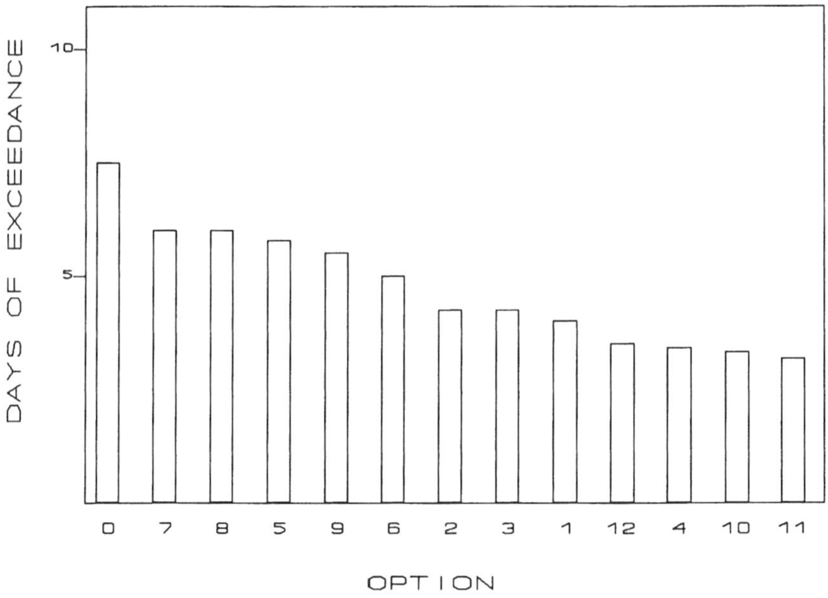

Figure 3–28. Monte Carlo simulations: Comparison of CSO control options in Poughkeepsie, New York, in terms of May-October average days of incremental fecal-coliform concentrations exceeding 1,000 counts/100 ml (Limno-Tech, Inc., August 1983).

40–180 times per year (depending on the degree of lateral mixing). Dissolved-oxygen impacts from CSO were small, although some poststorm reductions were observed due to reductions in sunlight and algal growth. Control options were evaluated and recommended using these comparisons as well as other engineering and economic considerations.

Onondaga Lake, New York

Syracuse, New York, is located on the southeastern shore of Onondaga Lake (Figure 3–29). A medium-sized metropolitan area with a population of approximately 136,000, Syracuse is served by 6,000 acres of combined sewers and 8,000 acres of separate sewers. Onondaga Lake had a long history of deteriorating water-quality but with significant improvement in the late 1970s. As part of a comprehensive program to monitor, study, and improve water-quality, Onondaga County undertook a CSO program to reduce related impacts (O'Brien and Gere 1979). The lake had a history of wet-weather depressions in dissolved oxygen and increases in fecal coliform. These problems, plus excessive eutrophication, restricted recreational use and aquatic health in the lake.

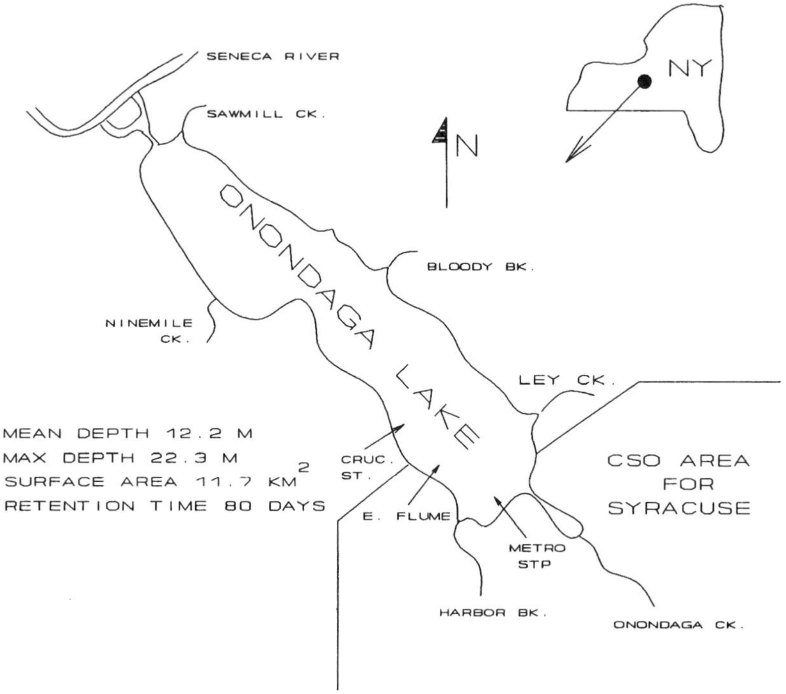

Figure 3-29. Onondaga Lake, New York, study area.

In addition to routine monitoring of baseline water quality in the lake, five intensive wet-weather storm surveys were conducted. These surveys included prestorm data, plus sampling up to five days following the storm event at 10 lake stations and seven tributaries. Parameters sampled focused on BOD, total Kjeldahl nitrogen (TKN), solids, dissolved oxygen, algae, fecal coliform bacteria, and streptococcus bacteria, with secondary emphasis on other nutrients, metals, dissolved solids, and conductivity. Each survey also included current measurements to define lake circulation.

Simulation of water quality in the lake required a uniquely developed model because of the three-dimensional aspects of pollutant transport (Limno-Tech, Inc. 1978). A 27-cell, 2-layered model was developed. Each cell was considered well mixed with advective and dispersive transport to adjacent cells. Model kinetics included dissolved-oxygen relationships, algal dynamics, and bacterial die-off. The model was calibrated and verified to each of the storms and then used to analyze the frequency and severity of problems under different conditions and CSO controls. Initial sensitivity showed that BOD and TKN loadings from CSO had little effect on poststorm dissolved-oxygen levels. Reductions were due instead to shading from clouds and storms that reduced dissolved oxygen from algal productivity. CSO loads had a dramatic impact on fecal coliform as shown in Figure 3–30. For example, for the once-in-one-year storm, the concentrations at the north end of the lake (farthest from the CSO) still peaked at over 5,000 counts/100 ml, and it took 2–3 days before levels dropped below swimming standards. Alternatives for controlling the CSOs were compared in terms of their effectiveness to reduce these impacts. This model has been subsequently modified to be more useable on a PC. Engineers involved in facility planning have interfaced the SWMM model with the bacteria model to project

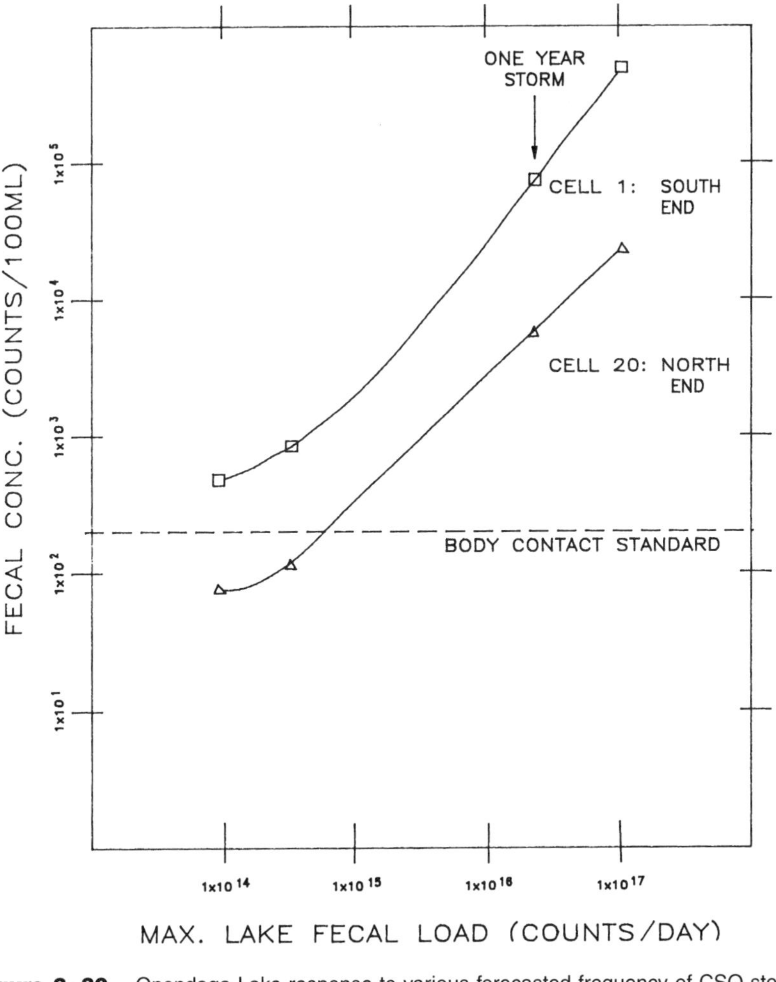

Figure 3–30. Onondaga Lake response to various forecasted frequency of CSO storm events (Limno-Tech, Inc. 1978).

improvements to Onondaga Lake over a 15-year schedule of improvements as described in chapter 5.

Perrysburg and Maumee, Ohio

Perrysburg and Maumee, Ohio, are two small suburban communities south of Toledo on opposite sides of the Maumee River (Figure 3–31). Combined population is approximately 27,000, with approximately 1,200 acres of combined sewered system (Finkbeiner, Pettis, and Strout, Ltd. 1982a, b). The Maumee River is free flowing but somewhat affected by the fluctuating level of Lake Erie. It has a mean flow of 4,844 cfs. Both cities had a frequent history of system

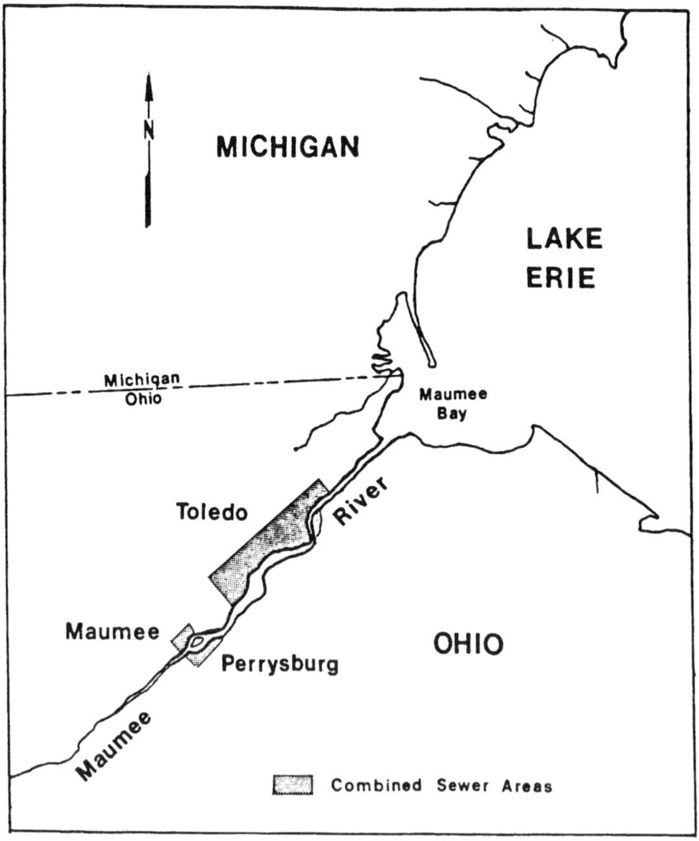

Figure 3–31. Location of the cities of Perrysburg and Maumee, Ohio (Limno-Tech, Inc., July 1982).

overflows even during small storms. For example, 80% of the storms in 1981 caused overflows.

Two dry-weather and three wet-weather water-quality surveys were conducted to document trends in water-quality and support-model development (Limno-Tech, Inc. July 1982). In addition, a continuous, dissolved-oxygen monitor was maintained to document the frequency of problems and baseline conditions. The intensive wet-weather surveys included 16 stations over an approximate 12-mile distance, with monitoring from 3 to 7 days depending on the poststorm recovery. The focus was on BOD, TKN, dissolved oxygen, and fecal-coliform impacts, but some composite samples were taken for other parameters. Surveys also included companion monitoring (and modeling) of CSO and wastewater-treatment-plant loadings.

The receiving-water-quality simulation used desktop analytical models. Dissolved oxygen was simulated using the one-dimensional advective/dispersive equation for pulse loading described previously. This framework allowed calculation of spatial and temporal profiles of dissolved oxygen in response to intermittent CSO loads. Fecal coliform was simulated using simple dilution calculations of river flow mixed with CSO inputs. For fecal coliform, only peak concentrations were of interest, and the dilution calculations were adequate. The receiving-

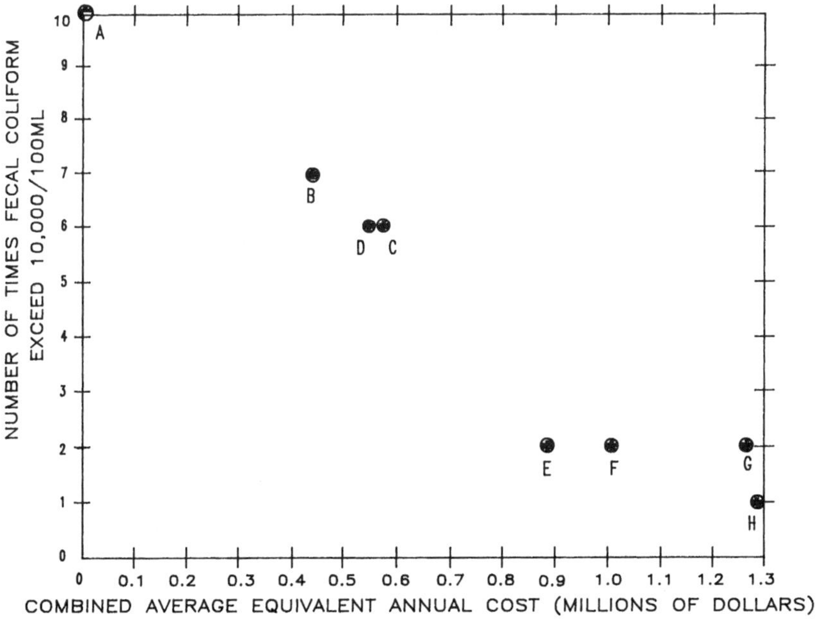

Figure 3-32. Model projected fecal-coliform levels above 10,000 counts/100 ml for Perrysburg and Maumee abatement combinations (Limno-Tech, Inc. 1982).

water models were linked to CSO simulation models for each city in order to simulate impacts for a typical year of storms, considering 11 and 17 alternatives for the two cities, respectively. At existing conditions, there were over 40 events per year where bacterial contact standards were predicted to be violated. Figure 3–32 shows how these violations could be reduced by seven different CSO control strategies involving different levels of satellite treatment, storage, and sewer separation. (In contrast, dissolved oxygen impacts were negligible.) Each alternative was evaluated by the reduction in days of violations in the bacterial-water, contact-recreation standards. Controls were selected based on cost and reduction in bacterial exceedances.

References: Chapter 3

Akar, P. J., and G. H. Jirka. 1991. CORMIX2: "An Expert System for Hydrodynamic Mixing Zone Analysis of Conventional and Toxic Submerged Multiport Diffuser Discharges," U.S. EPA Office of Research and Development, Environmental Research Laboratory, Athens, GA.

American Petroleum Institute (API). 1980. "Guide to Water Quality Standards of the United States," Environmental Affairs Dept., Washington, D.C.

Association of Metropolitan Sewerage Agencies (AMSA). January 1996. "Performance Measures for the National CSO Control Program," Washington, D.C.

Chapra and Reckhow. 1983. "Engineering Approaches for Lake Management, Volume 2: Mechanistic Modeling," Butterworth Publishing.

Dillon, J. J., S. Girman, and J. Marr. 1993. "Comparing CSO Control Costs to River Water Quality Impacts," WEF CSO Specialty Conference, Louisville, KY.

Doneker, R. L., and G. H. Jirka. 1990. "An Expert System for Hydrodynamic Mixing Zone Analysis of Conventional and Toxic Submerged Single Port Discharges," U.S. EPA Office of Research and Development, Environmental Research Laboratory, Athens, GA. EPA 600/3–90/012.

Effler, S. W., S. D. Field, M. A. Meyer, and Philip Sze. February 1981. "Response of Onondaga Lake to Restoration Efforts," ASCE, vol. 107, no. EE1.

Environmental Protection Agency. 1973. "Auto-Qual Modeling System," EPA 440 9–73 003.

———. 1977. "Guidelines for the Pollutional Classification of Great Lakes Harbor Sediments," Great Lakes National Program Office, Region V, Chicago, IL.

———. 1979. "User's Manual for the Dynamic (Potomac) Estuary Model Technical Report 63," EPA 903 9-79 001.

———. 1985. "Initial Mixing Characteristics of Municipal Ocean Dischargers: Vol. 1, Procedures and Applications," EPA 600 3–85 073a.

———. 1986. "Quality Criteria for Water 1986," EPA 440 5–86 001.

———. 1987. "The Enhanced Stream Water Quality Models QUAL2E and QUAL2E-UNCAS: Documentation and User Model," EPA 600 3–87 007.

———. 1988. "WASP4, A Hydrodynamic and Water Quality Model—Model Theory, User's Manual, and Programmer's Guide," EPA 600 3–87 039.

———. 1991. "Technical Support Document for Water Quality Based Toxics Control," Office of Water, Washington, D.C. EPA/505/2-90-001.

———. September 1993. "Technical Basis for Deriving Sediment Quality Criteria for Nonionic Organic Contaminants for the Protection of Benthic Organisms by Using Equilibrium Partitioning," Office of Water, Washington, D.C., EPA-822-R-93-011.

Finkbeiner, Pettis, and Strout, Ltd. 1982a. "Combined Sewer Overflow Study," prepared for City of Perrysburg, OH.

———. 1982b. "Combined Sewer Overflow Study," Sewer System Evaluation Survey-State No. 4, prepared for city of Maumee, Lucas County, OH.

Fischer. 1979. "Mixing in Inland and Coastal Waters," Academic Press.

Geldrich, E. E. 1966. "Sanitary Significance of Fecal Coliforms in the Environment," U.S. Dept. of the Interior, Publication no. WP–20–3.

———. 1978. "Bacterial Populations and Indicator Concepts in Feces, Sewage, Stormwater and Solid Wastes in Indicators of Viruses in Water and Food," G. Berg, ed. Ann Arbor Science Publications, Ann Arbor, MI. Pp. 51–97.

Greeley and Hansen; Limno-Tech, Inc.; and Hayes, Seay, Mattern, and Mattern. 1987. "Development of Preliminary Alternatives and Selection of Final Alternatives, Interim Report IV," prepared for city of Richmond, VA.

Hydroscience. 1979. "A Statistical Method for Assessment of Urban Stormwater Loads-Impacts-Controls," prepared for EPA, Non-Point Source Branch, Washington, D.C.

Jones, G. R., and G. H. Jirka. 1991. CORMIX3: "An Expert System for the Analysis and Prediction of Buoyant Surface Discharges—Technical Report," DeFrees Hydraulics Laboratory, School of Civil and Environmental Engineering, Cornell University, Ithaca, NY.

King County Metropolitan Services. 1996. Personal communications.

Limno-Tech, Inc. (LTI). January 1978. "Mathematical Modeling of the Impact of Storms on Water Quality in Onondaga Lake." Prepared for Stearns and Wheler, Syracuse, NY, and Onondaga County, NY.

———. July 1982. "Evaluation of the Need and Best Means for Combined Sewer Overflow Control for the Cities of Perrysburg and Maumee." Prepared for Finkbeiner, Pettis, and Strout, Ltd., Toledo, OH, and the cities of Perrysburg and Maumee, OH.

———. September 1982. "Comparison of Water Quality Impacts for District of Columbia Combined Sewer Overflow Control Alternatives." Prepared for O'Brien and Gere, Washington, D.C., and the District of Columbia, Dept. of Environmental Services, Washington, D.C.

———. August 1983. "Evaluation of the Need and Best Means for Combined Sewer Overflow Control for the City of Poughkeepsie." Prepared for O'Brien and Gere, Syracuse, NY, and the city of Poughkeepsie, NY.

———. 1987. "Assessment of Existing CSO Related Conditions in the Upper James River Estuary, Interim Report III." Directed by Greeley and Hansen, prepared for the city of Richmond, VA.

———. 1992. "Tidal Effect Correction Factor for the Willamette River, Portland, Oregon, PULSEQUAL Model Enhancements to Account for Tidal Effects." Prepared for the city of Portland, Bureau of Environmental Services, Portland, OR.

———. July 1992a. "Calibration of the PULSEQUAL Combined Sewer Overflow (CSO) Impact Model for the St. Joseph River, Study Memorandum #4." Prepared by LTI for the City of South Bend, IN.

———. July 1992b. "Water Quality Forecasts in the St. Joseph River, Based on 10-Year Continuous Simulations (1981–1990)." Prepared by LTI for the City of South Bend, IN.

———. 1993. "PULSEQUAL User's Manual." Prepared by LTI for the Water Quality Modeling Workshop conducted for the City of Portland staff between September 30, 1993, and October 1, 1993.

———. 1993. "Technical Memorandum 5.4-PULSEQUAL Model Calibration and Application for the Lower Willamette River at Portland, Oregon." Prepared for the Portland BES, Portland, OR.

———. 1995. "Development of a Tidal Version of the CORMIX Models for Application to Discharges in the Delaware Estuary—Final report." Prepared for Delaware River Basin Commission, West Trenton, NJ.

Milne, Ian, Edwin E. Hericks, and Ian Johnson. 1995. "Protocols for Wet-Weather Toxicity Testing and Assessment—Draft." Water Environment Research Foundation, Project 92–BAR–1.

NCASI, 1992. "User's Guide for the Cornell Mixing Zone Expert System (CORMIX)." NCASI Technical Bulletin No. 624.

O'Brien and Gere. 1979. "Final Report—Combined Sewer Overflow Abatement Program." Prepared for Onondaga County, Dept. of Drainage and Sanitation.

———. 1982. "Combined Sewer Overflow Abatement Program." Prepared for District of Columbia, Dept. of Environmental Services.

———. March 1983. "Final Report—Combined Sewer Overflow Study, Phase I." Prepared for city of Poughkeepsie, Dutchess County, NY.

Raytheon Company. 1974. "Receive-II Water Quality and Quality Model," Portsmouth, RI. Prepared for Office of Water Programs, EPA.

Scott, W. J. 1934. "Survey of Connecticut's Shore Bathing Waters," *Public Health Engineer* 19: 316.

Stevenson, A. H. 1953. "Studies of Bathing Water Quality and Health," *Journal of American Public Health Assoc.*, vol. 43: 529.

Thomann, R. V. 1972. *Systems Analysis and W.Q. Management.* New York: McGraw-Hill.

Thomann, R. V., and J. A. Mueller. 1987. *Principals of Surface Water Quality Modeling and Control.* New York: Harper and Row.

U.S. Dept. of the Interior. 1968. "Water Quality Criteria," National Technology Advisory Committee Report to the Federal Water Pollution Control Administration.

U.S. Environmental Protection Agency (EPA). 1984. "Technical Guidance Manual for the Regulations Promulgated Pursuant to Section 301(g) of the Clean Water Act of 1977," Appendix F.

———. 1985a. "Initial Mixing Characteristics of Municipal Ocean Discharges: Volume 1, Procedures and Applications." Office of Research and Development, Newport, OR. EPA 600/3–85/073a.

———. 1985b. "Water Quality Assessment: A Screening Procedure for Toxic and Conventional Pollutants in Surface and Groundwater—Part II (revised)," Environmental Research Laboratory, Athens, GA. EPA 600/6–85/002b.

Chapter 4

Control and Treatment of Combined Sewer Overflows

■ Richard Field, P.E., and Thomas P. O'Connor
Wet-Weather Flow Pollution Control Program
National Risk Management Research Laboratory
U.S. Environmental Protection Agency
Edison, New Jersey

Introduction

This chapter is a survey of control and treatment of combined sewer overflows (CSO), encompassing the U.S. Environmental Protection Agency (EPA), National Risk Management Research Laboratory, Wet-Weather Flow Pollution Control Programs, and other research efforts over the last 27 years. It divides control and treatment technologies into four sections:

(1) Collection system
(2) Storage
(3) Treatment
(4) Optimization

There is some overlap from section to section (e.g., storage in the sewer lines may be thought of as part of collection-system design, while sedimentation in overflowing storage tanks may be thought of as treatment; however, both topics will be examined in the storage section). Many of the control/treatment technologies and strategies presented here are also applicable to other forms of wet-weather flow (WWF) (i.e., sanitary-sewer overflows [SSO] and stormwater discharges).

The last section, optimization, will bring together the most efficient use in planning and implementing these storage, treatment, and collection system technologies. The CSO pollution abatement system must integrate storage and treatment because the storm flow is intermittent and highly variable in pollutant concentration and flowrate. Optimization begins with operational changes that increase the use of the existing system. Long-term urban-hydrological and CSO quality monitoring data should then be used to model the existing system, including treatment by sedimentation, in the existing storage facilities. Analyses should then be made of comparatively lower-cost modifications to or retrofitting of the collection system followed by higher cost modifications to the wastewater treatment plant (WWTP) (i.e., retrofitting the existing settling tanks). The more expensive options of constructing additional upgraded or parallel interceptors, storage facilities, treatment at the WWTP, and satellite treatment plants at the outfalls should be considered last.

Estimated capital and operation and maintenance costs from previous reports and papers that appear in this chapter have been adjusted to an *Engineering News Record (ENR)* index of 5000, except as noted.

Collection System

Improving Interceptor Capacity

The flow-carrying capacity of the interceptors can be increased by:

- cleaning out clogged and silted-up interceptors
- increasing the pumping capacity for surcharged interceptors
- installing self-cleaning systems (flushing)
- installing larger or parallel interceptors
- applying polymer injection or lining to reduce pipe roughness.

The intent of increasing the flow-carrying capacity of interceptors is to lessen the CSO pollution load. Taking these steps will also address several of the EPA's Nine Minimum Controls (EPA 1995a), specifically:

- proper operation and regular maintenance programs for the sewer system
- maximum use of the collection system for storage
- maximization of flow to the publically-owned-treatment works (POTW) for treatment
- control of solid and floatable materials in CSOs and
- pollution prevention programs to reduce contaminants.

The inspection of interceptors for blockages and siltation and the associated cleanout should be done first, since this is apt to be the least expensive way of gaining flow capacity.

Polymer Injection

Research has shown that polymer injection can increase flow capacity as much as 2.4 times at a constant head (The Western Company 1969b). This method can be used as a short- or long-term measure to correct troublesome pollution- and nuisance-causing conditions (e.g., localized flooding and SSO). A cost comparison in Garland, Texas, indicated that polymer injection for SSO control would cost one-fourth as much as relief sewer construction; however, additional cost verification is necessary for other locations (Field and Struzeski 1972).

Polymer injection was demonstrated in Dallas, Texas (Chandler and Lewis 1977). The existing sewer lines were relatively inaccessible but capable of handling normal DWF. An ongoing infiltration/inflow abatement program was already initiated; however, a quick, low-cost solution was needed to eliminate SSO in the interim. In the process of designing a full-scale automated injection system, portable and manned injection systems and the use of either dry or slurry polymer were demonstrated. Constant feed of polymer operated better than variable feed during WWF. Polymer injection was demonstrated to reduce or prevent SSO.

Sewer Flushing

Sewage solids deposited in combined sewer systems during dry weather is a major contributor to the CSO pollution load. The magnitude of these deposits that are flushed out by storm flows (WWF) has been estimated to range up to 30% of the total annual dry-weather-sewage loadings. Deposited solids can also rob the interceptor's flow-carrying capacity and, accordingly, cause premature CSO.

Periodic sewer flushing during dry weather is designed to remove the accumulating material and hydraulically convey it to the dry-weather flow (DWF) treatment facilities. This reduces the amount of solids resuspended during storm flow events, lessens the need for CSO treatment and sludge removal at downstream storage facilities, and could, if applied to interceptors, allow the passage of more WWF to the WWTP for treatment. Flushing is particularly beneficial for sewers with grades too flat to be self-cleansing. Sewer flushing requires cooperation between the authorities with jurisdiction over collection system maintenance and wastewater treatment.

As part of a sewer-flushing program it is necessary to estimate the deposition build-up. Field studies in Boston (Pisano et al. 1979) related the total daily mass of pollutant deposition in a collection system to the system characteristics (i.e., per capita waste production rate, service area, total pipe length, average pipe slope, and average pipe diameter). A simple model is given by the equation:

$$TS = 0.0076(L')^{1.063}\,(-S)^{-0.4375}(q)^{-0.51}\;(R^2 = 0.845)$$

where $TS \equiv$ deposited solids loading, lb/d

$-S \equiv$ mean pipe slope, ft/ft

$L' \equiv$ total length of sewer system, ft

$q \equiv$ per capita wastewater rate (plus allowance for infiltration), gal/capita/d

$R^2 \equiv$ coefficient of regression squared

Total pipe length of the system, L′, is generally assumed to be known. When this information is not known, the total pipe length can be estimated from the total basin area, A (acres), using these equations:

For low population density (10–20 people/acre)

$$L' = 168.95\,(\mathrm{A})^{0.928}\ (R^2 = 0.821)$$

For moderate-high population density (30–60 people/acre)

$$L' = 239.41\,(\mathrm{A})^{0.928}\ (R^2 = 0.821)$$

If data on pipe slope are not available, the mean pipe slope can be estimated using this equation:

$$\bar{S} = 0.348\,(\bar{S}_g)\ (R^2 = 0.96)$$

where $\bar{S}_g \equiv$ mean ground slope, ft/ft

Cleansing efficiency of periodic flush waves depends on flush volume, flush discharge rate, sewer slope, sewer length, sewer flowrate, sewer diameter, and population density. Maximum flushing rates at the downstream point are limited by the regulator/interceptor capacities prior to overflow. The relationship between cleaning efficiency and pipe length is important as the aim of flushing is to wash the resuspended sediment to strategic locations (i.e., a point where sewage is flowing, to another point where flushing will be initiated, or to the sewage treatment plant).

Sediment removal data as a function of length of pipe flushed (Pisano et al. 1979) are presented in Table 4–1. Estimated costs of sewer flushing methods are presented in Table 4–2.

Innovative methods for cleaning accumulated sludge and debris in CSO and stormwater conveyance systems and storage/sedimentation tanks have recently emerged. These methods create high-speed flushing waves to resuspend deposited sediments. The Hydroself and the Hydrass are two methods that use flushing gates.

Table 4-1. Pollutant Removals by Sewer Flushing as a Function of Length of Segment Flushed (254–381 mm [10–15 in.] pipe)

	% Removals: Organics and Nutrients	% Removals: Dry-Weather Grit/ Inorganic Material
Manhole-to-Manhole Segments	75–95	75
Serial Segments up to 213 m (700 ft)	65–75	55–65
Segment Lengths Greater than 305 m (1,000 ft)	35–45	18–25

The Hydroself system has been used in Europe for the last ten years. The system consists of a hydraulically operated flap gate, a flush-water storage area created by the erection of a concrete wall section, a float or pump to supply hydraulic pressure, and valves controlled by either a float system or an electronic control panel. The water level in the sewer or tank is used to activate the release and/or closure of the gate using a permanently sealed, float-controlled hydraulic system. Designed to operate automatically whenever the insystem water level reaches a predetermined level, the gate releases the stored water and causes a "dambreak" flushing wave to occur. Activation by remote control at the discretion of an operator is also possible. Depending on the physical constraints, the flushing system may have more than one gate.

The Hydroself flushing gate system has been used to flush and cleanse deposits and debris in troublesome sanitary, storm, and combined sewers and interceptors for both existing and new installations. For large diameter sewers (>6.5 feet [2.0 m]), the flushing system may be installed in the sewer pipe itself. The required storage volume for the flush water is created by erecting two walls in

Table 4-2. Estimated Costs of Sewer-Flushing Methods Based on Daily Flushing Program (ENR = 5,000) (Pisano et al. 1979)

Number of Segments:	46 (254–457 mm [10–18 in.] pipe)
Automatic-Flushing Module Operation (one module/segment)	
Capital Cost*	$21,640
Annual O&M† Cost	$199,084
Manual-Flushing Mode	
Capital Cost‡	$137,050
Annual O&M Cost	$236,738

*Includes site preparation and fabrication and installation of air-operated module.
†Operation and maintenance.
‡Includes three outfitted water tankers.
(EPA-600/2-79-133)

the sewer pipe, thus forming a flush-water storage area in between the two walls. For the area to remain free of debris, a floor slope of 20% should be provided in the storage area. The flushing length for a single flush gate is based on the storage area floor slope and available flushing volume. For longer flushing lengths, flushing gates may be installed to operate in series. In order to increase the effective flushing length, it is also possible to build additional flush-water storage area by creating a rectangular chamber inline or adjacent to the sewer line itself (Grande and Novac 1995).

Because the required flush-water volume cannot be built in the pipe for small diameter sewers, an inline storage chamber or one adjacent to the pipe may be installed. This system is demonstrated in Figure 4–1. In combined or sanitary sewers, a "flow-through-pipe," or underflow, is installed below the flush-water storage area to allow the DWF through the system unimpeded. When the inflow to the system exceeds the capacity of the underflow pipe, the water level rises and the storage area fills with flush water. Only after the sewer is drained or the peak flow subsides does the flushing system go into operation.

In Europe, 77 installations have been in operation since 1985 to flush sewers, interceptors, and tunnels ranging from 10 inches–14 feet (.25 m–4.3 m) in diameter and flushing lengths of up to 1100 feet (340 m) for larger diameter pipes and 0.62 miles (1000 m) for smaller diameter pipes.

A new flip gate flusher was developed in France called the Hydrass (Chebbo et al. 1995). This system uses a vertical gate pinned along a horizontal axis. The position of the axis causes automatic pivoting of the gate without mechanical assistance due to imbalance of hydrostatic head. When the pressure becomes greater on the portion of the gate above the axis, the gate pivots and releases a flushing wave downstream. When the storage upstream of the gate empties, gravity and water pressure reset the gate into vertical position to start another cycle. The filling time depends on the dry weather flowrate, the geometry of the sewer, and the height of the gate. This pivoting system has been installed on a segment of the Marseilles trunk sewer. Approximately 700 flushes were needed to clean a 330-foot (100 m) segment with an initial deposit of about 4 inches (100 mm). Flushing frequency can be reduced if the upstream elevation head can be increased as the flushing energy is more intense with greater heads (e.g., one flush at 4.9-ft (1.5 m) head is equivalent to 24 flushes at 1.6-ft (0.5 m) head).

Catchbasins/Catchbasin Cleansing

A catchbasin is a chamber or well usually built at the curbline of a street that admits surface water to a sewer or subdrain and has at its base a sediment sump. The sump makes the catchbasin more than a simple inlet device. Historically, catchbasins have minimized sewer clogging by trapping coarse debris from unpaved streets and have reduced odor emanations from low-velocity sewers by providing a water seal.

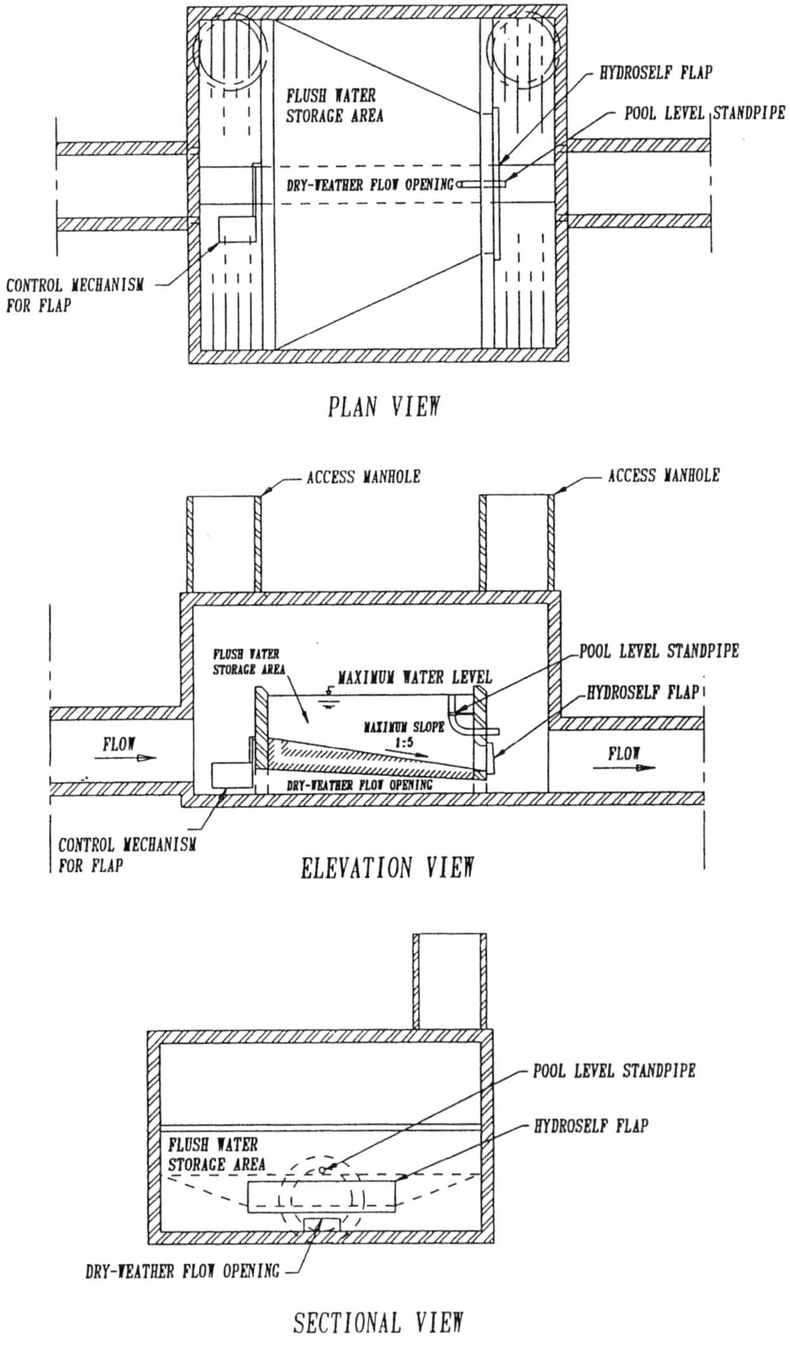

Figure 4–1. Typical Hydroself™ system installation for small-diameter sewers (Pisano).

Catchbasins must be cleaned often enough to prevent accumulated sediment and debris from blocking the outlet to the sewer. Many of the problems associated with catchbasins (i.e., blockages, odors, and pollution sources) are directly related to the inadequacies of the cleaning program. The sump must be kept clean to provide storage capacity for sediment and to prevent resuspension of sediment. Also, stormwater detained in the catchbasin, which eventually evaporates or leaks out, reduces the liquid storage capacity. In general, and depending upon local conditions, a cleaning frequency of at least twice per year maintains the effectiveness of catchbasins for pollutant removal. The increased cost of cleaning must be considered in assessing the practicality of catchbasins for pollution control.

In a project conducted in the West Roxbury section of Boston, Massachusetts, three catchbasins were cleaned, and, subsequently, four runoff events were monitored at each catchbasin (Aronson et al. 1983). Average pollutant removals per storm are shown in Table 4–3.

Typical installation costs for catchbasins range from $1,000–$2,500 and average $2,000 (Lager et al. 1977a). The reported costs will vary, depending on the size of the catchbasin. Catchbasin cost multiplication factors, as a function of sump storage capacity, are shown in Figure 4–2. Estimated national average costs for three catchbasin cleaning methods are presented in Table 4–4.

A recommended catchbasin configuration that evolved from hydraulic model studies (Lager et al. 1977a) is shown in Figure 4–3. The circular cross-section is preferred based on relative ease of cleaning and prefabrication. The dimension from the outlet pipe crown to the street or inlet grade is primarily a structural consideration, as it contributes little to the hydraulic performance. To be cost effective, the outlet pipe diameter (D_2 in Figure 4–3) should be sufficiently large to pass most flows under open-channel conditions, and a maximum allowable depth, a minimum of $4D_2$, should be used for the sump height.

Table 4–3. Pollutants Retained in Catchbasins (Aronson et al. 1983)

Constituent	% Retained
SS*	60–97
Volatile SS	48–97
COD†	10–56
BOD_5‡	54–88

(EPA-600/2-83/043)
*Suspended solids.
†Chemical oxygen demand.
‡5-day biochemical oxygen demand.

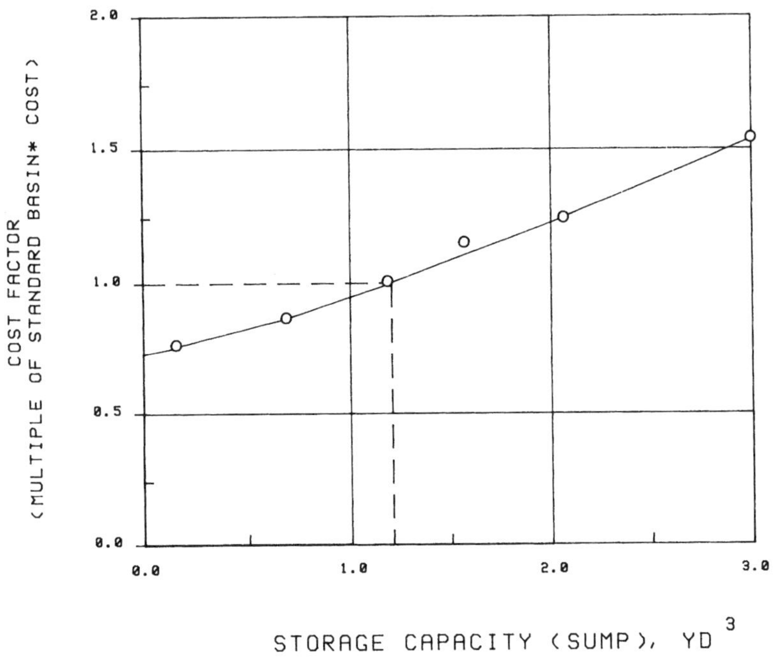

*The standard basin is basically a barrel 183 cm (6 ft) deep and 122 cm (4 ft) in diameter with an open top covered by a grating and an outlet pipe mounted at the side approximately 107 cm (3.5 ft) above the bottom.

Figure 4-2. Catchbasin cost factors versus storage capacity (Lager et al. 1977a).

Regulators

Controls used to restrict flow causing a backup and storage in the system fall into two categories: fixed or adjustable. Fixed systems are likely to be cheaper and require less maintenance but do not offer the flexibility and potential to maxi-

Table 4-4. Catchbasin Cleaning Costs (ENR = 5,000) (Lager et al. 1977a)

Manual			Eductor			Vacuum		
$/catch basin	$/m³	$/yd³	$/catch basin	$/m³	$/yd³	$/catch basin	$/m³	$/yd³
19.20	47.07	36.21	14.77	13.40	10.14	20.00	28.10	21.44

(EPA-600/2-77-051)

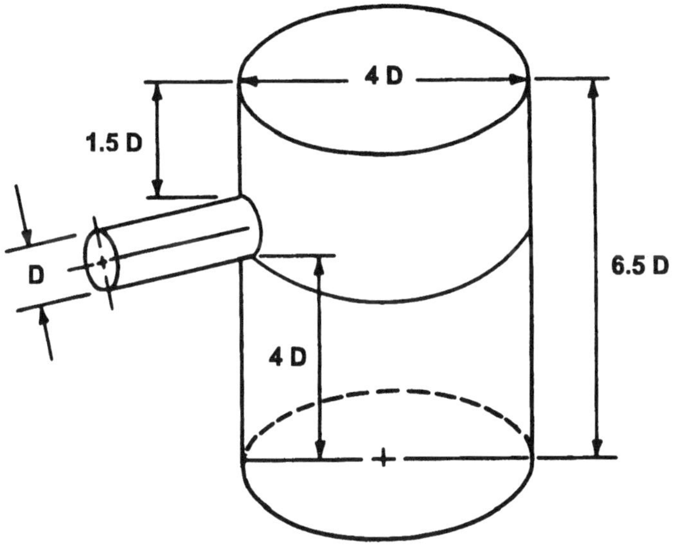

Figure 4–3. Recommended catchbasin design (Lager et al. 1977a).

mize the storage potential. Adjustable systems offer the advantage of being connected to a real-time control (RTC) system that, via a system of rainfall measurements and forecasts, monitoring of stormwater levels in critical sections of the drainage system, and input of these data into a computer system, can be adjusted to hold back or release stormwater to maximize storage capacity of the whole drainage system. Real-time control systems have been installed and are being further developed to control complex sewerage systems in the CSO field. The sophistication offered by an RTC system is unlikely to offer a cost-effective solution for a separate storm drainage system unless there is a large inline storage capacity and the stored runoff is to be treated.

Combined sewer systems are designed with CSO regulators to avoid hydraulic overload of sewers and WWTPs. For weir-type regulators, heights of the overflow crest are usually determined from the permissible backwater in the sewer system for a design flowrate and length of weir. Allowable weir heights affect the passive inline storage capacity. Long overflow weirs and corresponding large structures are often required for optimal use of the inline storage capacity of the sewer system.

Typical examples of fixed and adjustable flow regulators are:

Fixed Regulators	*Adjustable Regulators*
Orifices	Inflatable dams
Weirs (lateral and longitudinal)	Tilting plate regulators

Vortex flow throttle
Swirl/vortex/helical bend (separators)
Stilling-pond weir
Reverse-tainter gates
Float-controlled gates
Motor-operated or hydraulic (usually sluice-type) gates

Some of the preceding flow regulators are relatively inexpensive, quick to install, and effective at increasing storage. Several publications (Urbonas and Stahre 1993, Lager et al. 1977, American Public Works Association [APWA] 1970a and 1970b) on CSO control can provide more information on these regulators. However, as stated earlier, without treatment the advantage of storage is only in flow attenuation. It should also be noted that some of the regulators will concentrate the heavier solids into the underflow, which will be conveyed to the interceptor for treatment. This type of regulator (i.e., the swirl/vortex and helical bend regulators/concentrators) will be discussed later.

Innovative Regulators

Within the last two decades several innovative regulators have been developed and demonstrated.

Turbo

A throttle with adjustable feedback control not requiring an electrical source was developed by Brombach (1989a). A schematic of the Turbo regulating throttle is depicted in Figure 4–4. Flow enters the vortex throttle tangentially. In the upper section of the vortex chamber is a light, flat turbine wheel of synthetic material. A screw shaft leads to an oil pump through the hinged-dish head cover. The pump is supplied with hydraulic oil from a reservoir above the sluice gate. Since the vortex chamber is partially filled during DWF, the turbine wheel is dry and motionless. As the inflow increases during a storm, the swirling flow in the chamber engages the turbine wheel, causing the oil pump to push hydraulic oil into the oil actuator, which is housed in the head flange of the hydraulic cylinder. The actuator activates a low-pressure, hydraulic, constant-rate cylinder driving a moveable knife gate. This gate throttles the inflow to the vortex chamber just enough so that the turbine speed matches the preset design flow. Should the flow fall below the design flow, then the turbine wheel becomes still and a rebound spring opens the sluice gate.

The system is a true flow regulator with closed feedback loop since the vortex valve and turbine wheel serve simultaneously as a flow meter and source for the throttling energy. The dynamic behavior of the Turbo's regulating cycle was optimized using a combination of laboratory and mathematical simulation tech-

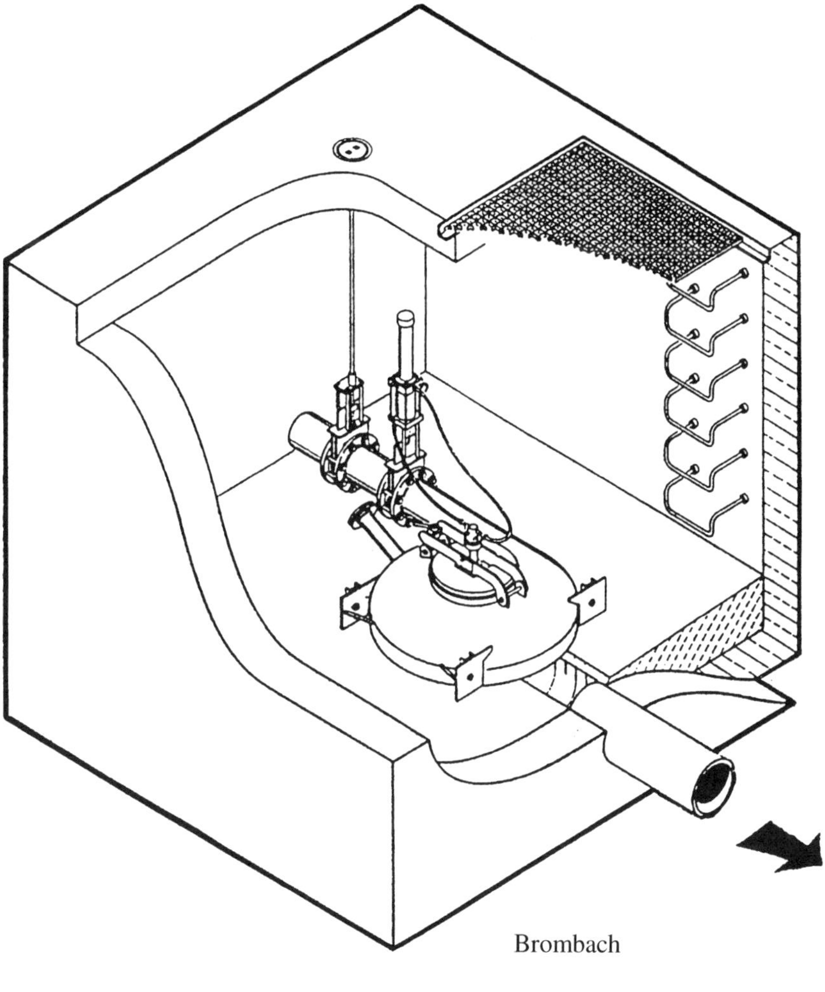

Figure 4-4. Turbo™ Vortex controller.

niques. The Turbo stabilizes remarkably in less than a minute even under extreme conditions (i.e., surges from zero to design-flow conditions). This rapid dynamic stability is essential for regenerating the regulating cycle during clogging. The system has unusual passive reserves. Should the oil hydraulic controller fail (e.g., from oil leakage loss), then the vortex valve still serves as an emergency brake. If the sluice gate hits a rock while closing, then a relief valve within the actuator allows the gate to rebound. All moveable parts remain in the oil stream. Jamming, freezing, or corrosion do not occur.

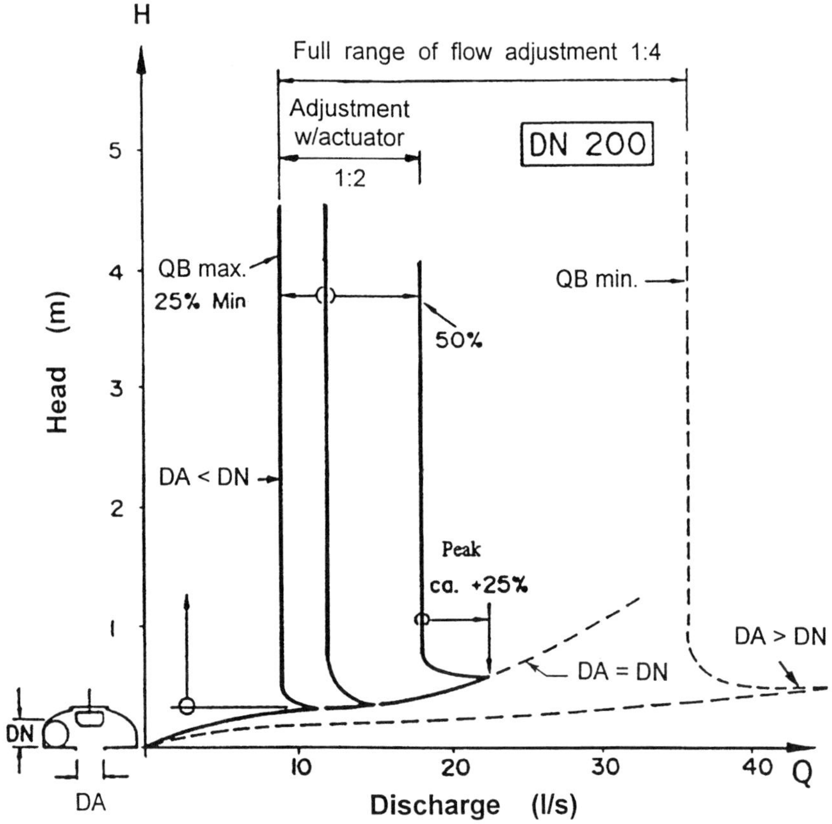

Figure 4–5. Discharge curves for Turbo™ Vortex.

During DWF, the Turbo flowrate curve is slightly S-shaped (see Figure 4–5) and is determined by the hydraulic operation of the vortex throttle. As the flowrate increases, the actuator and the sluice gate operation maintain a constant discharge independent of head. A nominal design flowrate can be adjusted in the internal bypass valve within the actuator. This change causes the turbine wheel to turn faster (or slower) to change the sluice gate position. Small flow adjustments can be accomplished in this manner (Figure 4–5). Larger flow adjustments can also be easily accomplished. In the center outlet of the vortex throttle is a loose plastic orifice plate. The Turbo design discharge can thus be modified in the ratio of 1:4 using a combination of both adjustments. The practical technology range of the Turbo is about 0.23–2.7 Mgal/d (860–10,400 m^3/d). Today, there are over 100 installations in Germany for controlling tank outflow using the Turbo.

Hydroslide

The Hydroslide flow regulator was developed in Germany. To date there are over 6,000 units in operation. The device uses a float-activated mechanism to maintain a constant discharge without the use of external energy sources. Figure 4–6 depicts the device in the neutral or fully open position in a typical installation during dry weather. The float mechanism is at its lowest position, thus providing the largest flow opening at the time when the regulator is most likely to experience a blockage. The device operates only when inflow is greater than the capacity of the fully open unit. When this occurs, the increasing water level causes the float to rise. This in turn causes the flow area to be adjusted so that a

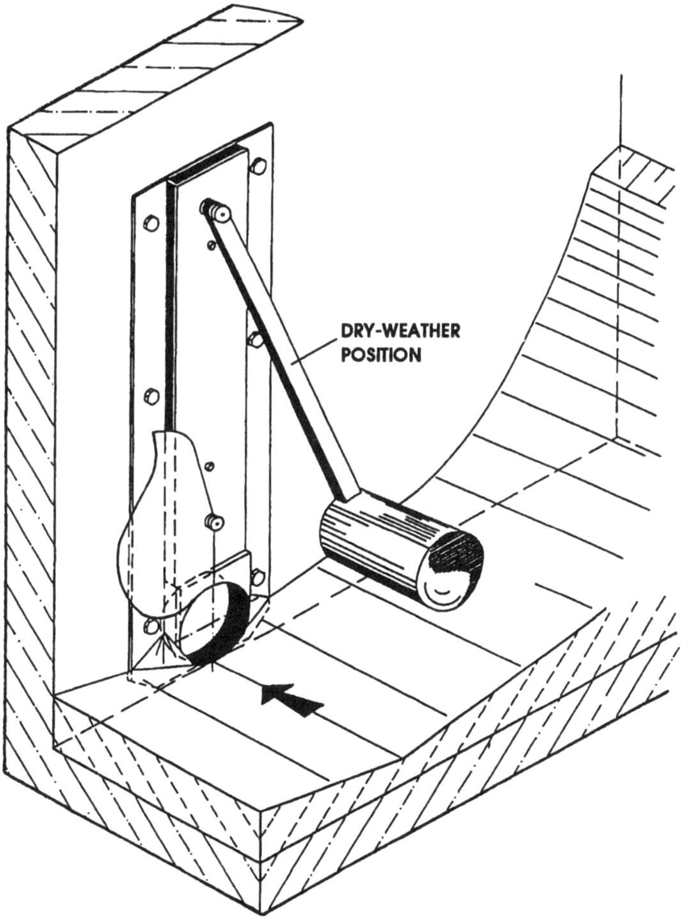

Figure 4–6. Hydroslide™ flow throttle (Adapted from Grande, Novac and Associates Catalog).

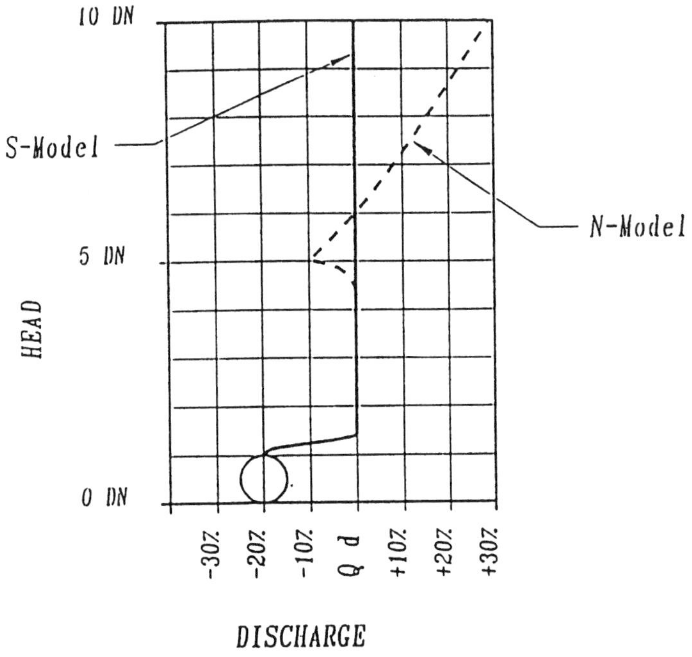

Figure 4–7. Typical discharge curves for Hydroslide™ flow regulator (models S and N).

constant discharge is maintained. Figure 4–7 depicts typical flowrate curves for Hydroslide flow regulator models N and S. The discharge increases rapidly as the upstream water level rises to achieve the design discharge at a head of approximately 1.4 times the unit nominal diameter, DN. For model S the flowrate is maintained constant for head variations of up to 12 times the nominal inlet diameter, while for the N model a constant discharge is maintained for 6 times DN (Novac et al. 1996).

Bending Weirs

Two different types of bending weirs have been developed in Germany that are more efficient than conventional fixed weirs (i.e., they are much shorter in length for comparable flowrate and backwater depth). Both operate automatically without the aid of auxiliary power. One is called the Fluidbend (Brombach 1992) and utilizes a very light and strong bending flap made of tempered high-strength spring steel. Both the shape of the flap and its metal properties determine the bending characteristic. The other—called the Hydrobend (GNA 1994)—creates the bending characteristic using a composite system of a stiff weir with moving bearings, counterweights, and axles.

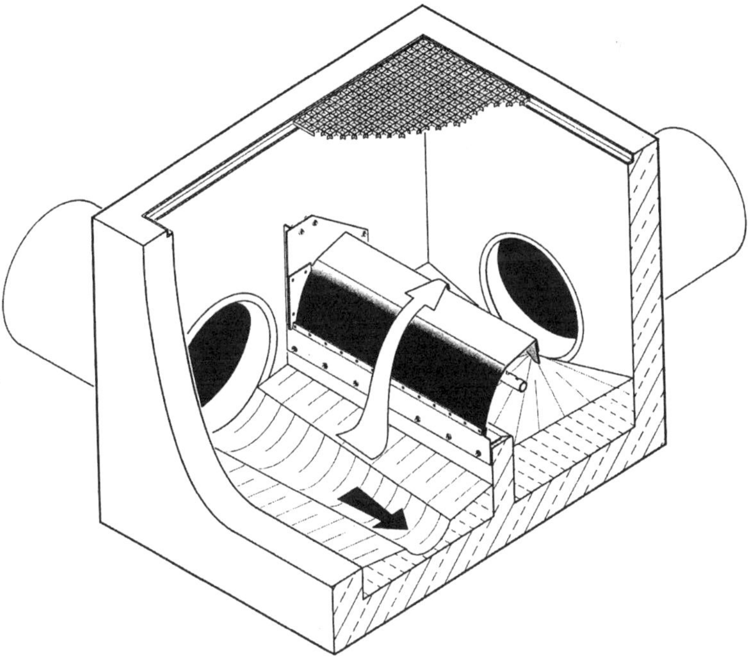

Figure 4–8. Fluidbend™ bending weir.

The Fluidbend is shown in Figure 4–8. The shape of the bending sheet and selection of materials are the result of extensive laboratory tests and structural calculations. Experiments have indicated a very complex relationship between the effects of static and dynamic forces and the passive restoring forces of the bend. Briefly, the metal weir is attached on one side and will bend under the force of increasing water level. A combination pipe stop and aeration tube prevents collapse. The weir is manufactured in different gage thicknesses (stiffness) to permit flexible design. Maximum rated discharge is about 4.9 Mgal/d/ft (60,500 m^3/d/m) under a maximum head of 10 inches (0.25 m). There are more than 30 inline storage control applications in Germany. The maximum bending weir span has been 10.6 feet (6.0 m). The device is intended for low-head weir installations (weir nape limited to 10 inches [0.25 m]).

The Hydrobend (Figure 4–9) is designed to permit overflow only when the maximum design water level behind the weir is reached. It will remain in the neutral (closed) position as long as the preset design maximum upstream water level is not achieved. Once the maximum storage capacity is achieved, the device reacts and adjusts itself so that the overflow discharge matches the additional inflow. The key component of the Hydrobend is an eccentric control disc that is designed to provide a balance of forces between the hydrodynamic pres-

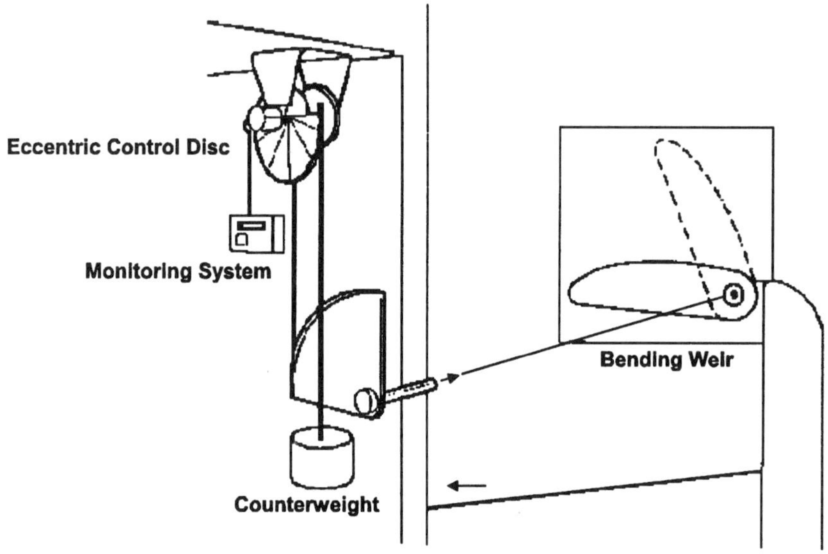

Figure 4-9. Hydrobend™ bending weir.

sure of the flowing water acting on the overflow face of the bending weir and the counterbalancing weight. Should the additional inflow be greater than the design peak discharge, the Hydrobend will remain fully open and act as an emergency overflow. The Hydrobend has been in use in Europe since 1985 with a total of 67 installations. Median design discharge and weir length and height for the 67 installations are 50 Mgal/d (190,000 m^3/d), 12 feet (3.6 m), and 2 feet (0.6 m), respectively.

Flow and level control are problems frequently encountered in the design of overflow discharge structures. Static overflow weirs with a fixed discharge level are generally chosen as they are simple to design and usually have predictable hydraulic characteristics. A static "long" weir is always designed using peak probable flow, which can result in long weir lengths and expensive structures. Mechanical gates are often used to overcome the limitations of static weirs (decrease length) but require instrumentation and energy (power), have numerous moving parts, and delicate seals (maintenance).

Siphons

Siphons in the form of spillway discharge have been used in water resource hydroelectric designs to rapidly spill large quantities of water where flow level control within a narrow head range is desired and available weir spillway space is limited. Maintenance of the siphon prime, however, has always been extremely difficult. Although the siphon spillway concept is an attractive alternative to both static long spill weirs and mechanically operated short weirs, the development of control measures to ensure siphon stability (mitigation of flow hysteresis [i.e., loss of prime]) has until recently precluded utilization of this concept.

Although the siphon has been known since the time of the Egyptians, the concept of regulating the flow through a siphon by controlled admission of air is a much later development. The recently developed air-regulated siphon (see Figure 4–10) (Markland and Brombach 1987) eliminates these technical problems and offers an attractive alternative to conventional systems. Stormwater overflows using the air-regulated siphon can be much smaller in size than conventional weirs. For existing structures, the air-regulated siphon provides opportunities for increasing overflow weir discharge to enhance the structure's capacity to handle infrequent flow peaks without surcharge. To date, over 40 devices have been installed, mostly in Germany.

The new air-regulated siphon has a vertical suction tube, a specially designed trumpet-shaped inlet, and a U-shaped discharge section. The device is very compact and can be installed on a horizontal base on the top of a standard static weir. The air inlet pipe is slightly inclined and connects the vertical discharge chute to the upper part of the siphon both inside and outside. This unique

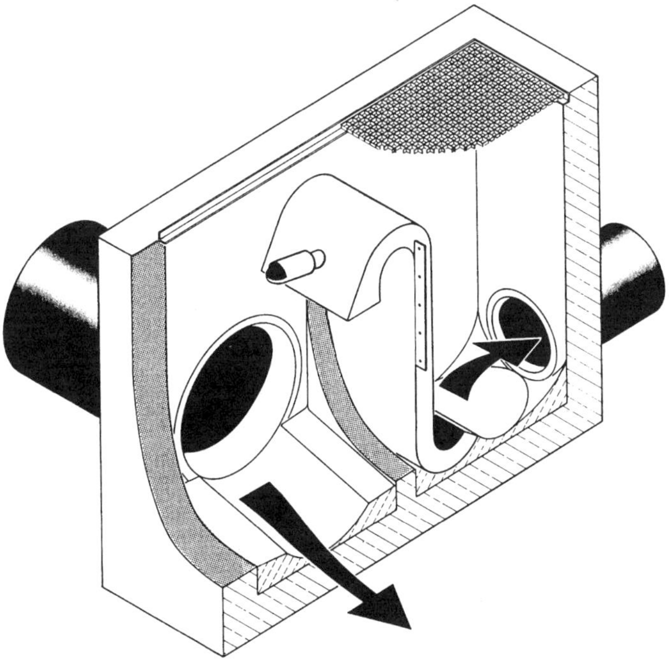

Figure 4–10. Air-regulated siphon.

air intake pipe has two orifices: The first is at the vertical discharge chute, and the second is in the interior section of the intake. The air inlet connects this tube with the interior of the siphon. A special "bump" creates a flow-deflecting effect that helps initially to establish the siphon. A small orifice located at the bottom of the discharge section permits drainage following an event.

When the water level rises to a level of 70% of the nominal siphon throat size, the air intake will be completely submerged, admitting only water. At this point the siphon is fully primed and will discharge at full or "design capacity." The siphon discharge capacity will increase only slightly with increasing water depth until the siphon itself is fully submerged. At this point a combination of weir-type discharge and siphon flow at full design capacity will occur. A typical design shows the merit of the device (Brombach 1986). Assume a siphon with a nominal throat of 7.9 inches (0.20 m), a weir width of 3.3 feet (1.0 m), and a suction head (recoverable head behind weir) of 6.6 feet (2.0 m). This configuration requires only 5.5 inches (0.14 m) of head to attain a peak flowrate of 16.2 Mgal/d (61,600 m^3/d). The equivalent rounded-edged broad crest weir of the same length operating under the same head would discharge only 2.32 Mgal/d (88,100 m^3/d). The attractiveness of this siphon is directly related to the available suction head that can be recovered. If the suction head were only 3.3 feet (1.0 m), then the design flowrate of the siphon would be 12 Mgal/d (45,000 m^3/d). As the available suction head decreases, the advantage diminishes.

Vortex Separators

Vortex solids separators are compact devices that provide flow regulation and a rough level of suspended solids (SS) and floatables removal. The technology was first applied in England and has been in use for over 30 years. Bernard Smisson first incorporated a cylindrical vortex-type CSO regulator/settleable-solids concentrator into the Bristol, England, sewerage system in the early 1960s. In the early 1970s, the EPA completed a series of projects to develop and demonstrate swirl settleable-solids removal technology for North American practice. These projects resulted in the EPA swirl CSO flow regulator/settleable-solids concentrator (hereafter referred to as swirl) and the EPA swirl degritter, a variation that does not regulate flow while imparting settleable-solids (or detritus) separation. Subsequently, new generations of vortex technology emerged, including the Fluidsep and the Storm King vortex-hydrodynamic separators. The differences among the three types came about through experimental work aimed at creating flow conditions with the unit that would optimize the liquid/solids separation. Each of the three devices is equipped with baffles and traps for capturing and storing floatables during an event. Despite design and application differences, the main intent of these technologies is the same: to separate settleable solids from the storm flow by a vortex or swirling flowfield.

Vortex solids separators are mostly used for settleable solids control, but floatables can also be removed. Definitive floatables removal data for the three devices are extremely limited. It is generally held through visual observation that floatables capture in vortex separators deteriorates with increasing hydraulic loadings. This is especially true for the neutrally buoyant floatables or "swimmers."

Swirl SS separation efficiency depends upon the design flowrate (Q_d) and the fraction and inertial characteristics of settleable solids in the total CSO or storm-flow influent. The swirl was developed through hydraulic modeling studies on a 3-ft (0.9 m) diameter swirl model with representative synthetic particles (primarily gilsonite with a specific gravity [SG] equal to 1.06 and effective diameters [d_e] from approximately 0.5–3 mm) (Sullivan et al. 1974). Using the Froude and Stokes laws, the synthetic particles and swirl model were scaled (1:12) to predict the removal of actual grit (fine sand) (SG = 2.65 and d_e from approximately 0.2–2 mm) and relatively heavy organics (SG = 1.2 and d_e from approximately 0.2–5.0 mm) for a 36-ft (11 m) diameter prototype swirl (Sullivan et al. 1972). The swirl's flow patterns are effective for the separation of only the relatively large d_e and high SG particles, and its separation effectiveness decreases as particle d_e and SG decrease. The laboratory test solids represent only the heavier, denser fraction of solids found in CSO. Actual CSO contains a wider range of solids, so removals in field operations are closer to 40–50% removal.

The following describes the swirl's operation and can be best understood by referring to Figure 4–11. Flow enters the swirl tangentially and follows the peripheral wall of the cylindrical swirl chamber creating a swirling vortex-flow pattern. The swirling action causes settleable solids to be concentrated at the bottom of the unit. These solids are then routed into the underflow, which exits through a foul-sewer outlet in the bottom of the unit. This underflow ranges from 6%–10% of influent at Q_d.

The clarified effluent exits the top via an annular overflow horizontal weir and weir-plate arrangement into the central downshaft. The weir-plate is baffled to impede the free vortex-flow pattern. Coarse floatable matter was also simulated for hydraulic-model development (Sullivan et al. 1972). Another baffle configuration outside the weir plate captures floatables in the supernatant by directing them under the weir plate for containment during the storm event. The floatables are then carried out in the underflow as the storm flow subsides and the water level in the swirl falls. During low- and dry-weather-flow conditions, all flow passes out via the bottom foul-sewer outlet.

The size of the swirl, specifically the inlet and chamber diameters (all other dimensions are geometrically proportional to these components), is based on Q_d and the desired removal efficiency for the settleable solids. Detailed design information for the swirl is contained in the EPA publication, "Design Manual: Swirl and Helical Bend Pollution Control Devices" (Sullivan et al. 1982).

If site constraints are encountered, the swirl affords some flexibility in its sizing. Depending on depth (i.e., difference in elevation between the trunk-

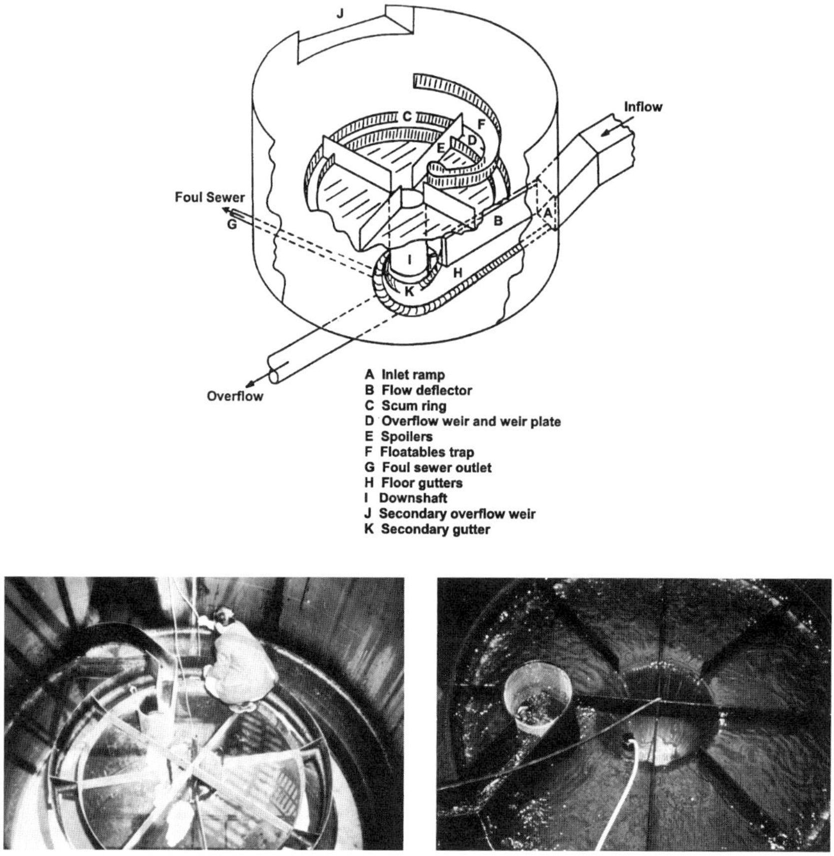

Figure 4-11. Swirl isometric (Sullivan et al. 1982).

combined sewer and the interceptor) and site area availability, respectively, the height to diameter configuration of the swirl can be adjusted by using a greater diameter for a lesser height and vice versa (Sullivan et al. 1974, 1982).

Design criteria for the Fluidsep were developed via hydraulic model testing (Brombach 1987a, 1987b). Capture efficiency of nine different synthetic tracers covering the full range of grit material with settling velocities ranging from 0.1–15 cm/sec was noted for a range of hydraulic loadings for three vessels having different diameter-to-depth ratios. Floatables tracers were also used (−2−−8 cm/sec) in the experiments. Threshold hydraulic limits were determined as to when floatables could not be captured or retained and expelled from the unit. These experiments indicated that settleable solids could still be removed even when this flow threshold was exceeded (Pisano 1990). Substantial floatables (both risers and swimmers with slight buoyancy) loss has been visually noted after storm events (effluent passing through coarse horizontal screens) at several Fluidsep installations (Tengen, Germany, and Burlington Vermont) when the influent hydraulic loadings exceeded 57 gal/ft^2/min (140 m^3/m^2/h).

The Storm King design is based on pilot-scale testing at the site. Figures 4–12 and 4–13 present generalized drawings of the Fluidsep and Storm King units, respectively. Pilot-scale testing is the optimal way of determining whether the swirl is suited for a particular site. Although this increases preliminary design costs, it represents only a fraction of the overall construction costs. A pilot-scale swirl can be set up by diverting a side stream from the storm-flow mainstream.

Swirl placement affects its pollutant removal effectiveness. Because swirls were developed to remove the relatively heavy particles, they should not be placed downstream of storage/sedimentation basins or grit chambers. The swirl, when properly placed and applied in the combined sewerage or storm-drainage

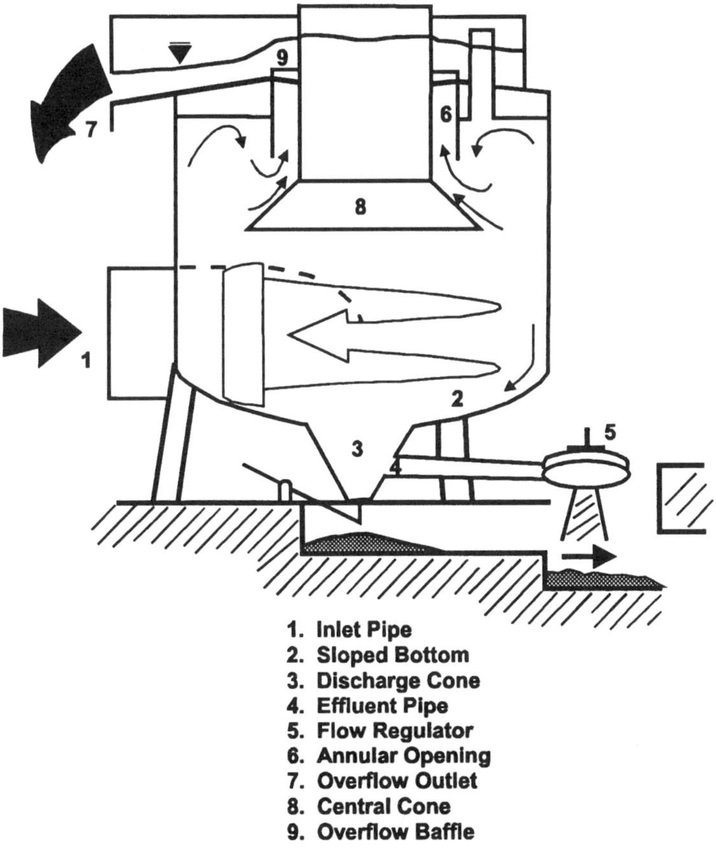

Figure 4–12. Fluidsep™ isometric.

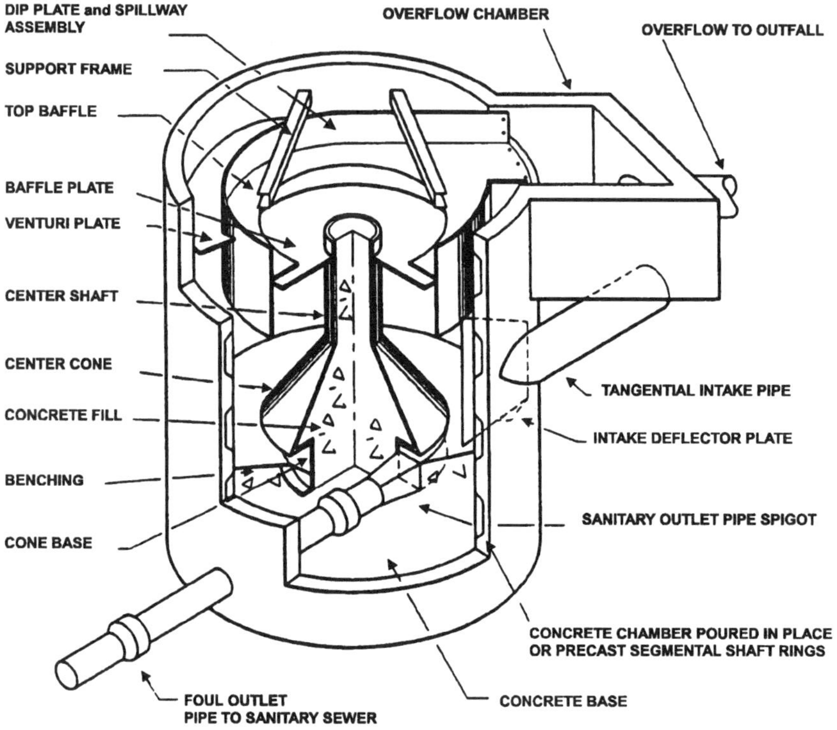

Figure 4-13. Storm King™ isometric (HIL Technology, Inc., 1991).

system, will be an effective tool for pollution-abatement optimization. Effective placement or application of the swirl includes:

- location upstream of a storage/sedimentation facility. The swirl effluent is stored for eventual treatment at the WWTP, or, in cases of excessive storm flows, the effluent receives additional sedimentation treatment in the storage facility before it is discharged to the receiving water. The swirl will then reduce the operation and maintenance necessary for removing grit and sludge from the storage facility as the heavier settleable solids are diverted directly to the WWTP, especially upstream of deep tunnels to relieve the burden of grit and sludge removal.
- use as a flow regulator/settleable-solids concentrator alone with the relatively clear effluent discharging directly to the receiving water. This application should be used only in cases where the receiving water can assimilate a partially treated storm-flow discharge and still meet its designated water uses.

- use of the swirl degritter when coarse or preliminary treatment is the only objective and flow regulation or flow splitting is not required. In this case, detritus (grit and heavy organic particles) is collected in a bottom hopper for later removal.
- placement of the swirl degritter in series with a swirl to remove detritus from the foul-sewer underflow (especially in cases where the underflow has to be pumped back to the WWTP) as was done in the Lancaster, Pennsylvania, project (Pisano et al. 1984).
- downstream of inline or in-sewer storage, which was demonstrated for the Northeast Boundary, Washington, D.C. project (O'Brien and Gere Engineers, Inc. 1992).

The Storm King and Fluidsep have been installed offline with the primary function as treatment devices. Potential underflow pumping requirements increase operation and maintenance costs (especially for the Storm King and Fluidsep, which have higher elevation-head requirements than the swirl).

The fraction of the influent SS (and associated pollutants) that are concentrated by the swirl into the underflow indicates the portion of SS diverted from the effluent (overflow) to the WWTP. The three performance indicators for swirl and vortex treatment are:

- Removal
- Net Removal
- Treatment Factor (TF)

These performance indicators are defined by the terms in these storm-flow-event summation equations (development from Field and O'Connor 1996):

$$M = \sum_{j=1}^{n} c(\xi_j)Q(\xi_j)\Delta_j t$$

$$V = \sum_{j=1}^{n} Q(\xi_j)\Delta_j t$$

$$C = \frac{M}{V}$$

where $M \equiv$ storm-flow-event pollutant mass loading (mass), $V \equiv$ storm-flow-event flow volume (volume), $C \equiv$ storm-flow-event, flowrate-weighted-average pollutant concentration (flow-weighted-average concentration), $c(\xi_j) \equiv$ average pollutant concentration between samples, $Q(\xi_j) \equiv$ average flowrate between samples, and $\Delta_j t \equiv$ time interval between samples.

These terms can be combined to form the equation for removal.

$$\text{Removal} = \frac{C_i V_i - C_e V_e}{C_i V_i} \times 100\% = \frac{M_i - M_e}{M_i} \times 100\%$$

where $M_i = C_i V_i \equiv$ mass of untreated influent and $M_e = C_e V_e \equiv$ mass of treated effluent.

For most treatment unit operations (e.g., settling, screening, and filtration), the volumes of the influent and effluent are equivalent and therefore may be canceled out of the removal equation, which can then be rewritten in terms of concentration alone. However, this approach, using only concentration to measure performance, cannot be used for the swirl because the swirl treatability evaluation is complicated by the continuous and relatively dilute underflow. The volume of the underflow remains significant throughout the event, and its percentage of the influent is highly variable (6%–10% at Q_d and as high as 100% for the smaller storms, which are completely captured in the swirl chamber without causing effluent). Therefore, a different approach is necessary to evaluate the concentrating effect of the swirl and other vortex units.

Assuming there is no net concentrating effect by the swirl or vortex (i.e., SS concentration of the influent, effluent, and underflow are all equal), then C would cancel out of the removal equation. The equation would then reflect only the flow-splitting nature of these devices and is termed the Reduction.

$$\text{Reduction} = \frac{V_i - V_e}{V_i} \times 100\%$$

where: $V_i \equiv$ influent volume and $V_e \equiv$ effluent volume. This nonconcentrating flow phenomenon is similar to what occurs during the operation of conventional CSO flow regulators. Accordingly, this can further be thought of as the CSO pollution reduction resulting from a conventional flow regulator.

Removal and reduction can now be combined to define the net removal and TF, which represent the pollution removal above and beyond the reduction gained by conventional CSO flow regulation. The equation for net removal is:

$$\text{Net Removal} = \text{Removal} - \text{Reduction}$$

A positive net removal indicates that SS (and associated pollutant) concentration has taken place in the unit. The TF equation is:

$$TF = \frac{\text{Removal}}{\text{Reduction}} = \frac{(C_i V_i - C_e V_e)/C_i V_i}{(V_i - V_e)/V_i} = \frac{C_u}{C_i}$$

where $C_i \equiv$ influent flow-weighted-average concentration and $C_u \equiv$ underflow flow-weighted-average concentration. *TF*s > 1 indicate that the vortex separator is concentrating SS to the underflow. The higher the *TF*, the better the vortex device concentrates pollutants. A negative net removal or a $TF < 1$ indicates an anomaly or faulty monitoring techniques.

These equations define the terms used in the tables presented in the following subsection. In the case of the swirl degritter and Grit King, again as with most forms of treatment, the influent volume can be treated as equivalent to the effluent volume.

Case Studies

The Northeast Boundary, Washington, D.C., swirl facility is comprised of three 57-ft (17.4 m) diameter swirls, each having a Q_d of 133 Mgal/d (503,000 m^3/d) with a hydraulic loading of 35 gal/ft^2/min (84 m^3/m^2/h). Samples were obtained with samplers having proper intake velocities and evaluated. Removal averaged 38% for all storm events; however, most of this removal (26%) was accomplished by direct foul-underflow diversion to the WWTP or reduction and not by the swirl concentration effect. Table 4–5 shows that net removal was

Table 4–5. Swirl Performance, Northeast Boundary CSO Facility, Washington, D.C. (O'Brien & Gere Engineers, Inc. 1992)

Storm 1	Removal (%) 2	Reduction (%) 3	Net Removal (%) 4	Treatment Factor (%) 5	Predicted Removal[a] (%) 6
1	28.3	14.6	13.7	1.94	16.6
3	67.5	64.4	3.1	1.05	67.0
4	39.4	33.9	5.6	1.16	—
5	27.9	14.9	13.0	1.87	22.2
7	42.0	24.3	17.8	1.73	26.0
8	35.6	10.8	24.8	3.30	10.5
9-SW2	48.9	27.1	21.8	1.83	—
10	34.9	23.9	11.0	1.46	—
11	30.0	30.0	0.0	1.00	25.7
12	21.1	12.6	8.5	1.67	32.1
Average[b]	37.6	25.7	11.9	1.46	28.6

Note: Design flowrate = 503,000 m^3/day, diameter = 17.4 m, and hydraulic loading = 1.4 m/min. Performance based on SS analyses.

[a]Based on settling-velocity analyses.

[b]Average of storm events.

less than expected, averaging 12%, with a TF of only 1.46. The primary cause of low net removal appears to have been the relatively low settling velocities of particles in the influent CSO. However, an important observation of this swirl's performance is that the theoretical (predicted) average removal, estimated to be 29% based on measured SS settling velocities (averaged for seven storm events), were lower than the actual measured removals of 38% (averaged for all ten storm events). The facility floatables traps were reported to be effective.

The Northeast Boundary, Washington, D.C., swirl facility also shows that the swirl is capable of removing particles lighter than those used for swirl development. For Storm 5, 50% of the influent SS had settling velocities greater than 0.1 cm/sec, and only 10% had settling velocities of 1.0 cm/sec or greater (O'Brien & Gere Engineers 1992). Thus, even with less than 10% of the influent SS concentration with settling velocities in the range of the settling velocities of the particles used for swirl development, a removal of 27.9%, a net removal of 13%, and a TF of 1.87 were achieved for Storm 5.

Evaluation of the 24-ft (7.3 m) swirl in Lancaster, PA (Pisano et al. 1984), showed that proper sampling will result in higher SS removals and net removals. The Lancaster, PA, demonstration project used a specially constructed cross-sectional sampler capable of automatically taking a vertical slice or segment of the whole influent conduit cross-section (Pisano et al. 1984). Prior to using the slice-sampler, the Lancaster swirl had shown "negligible to negative solids treatment efficiency" and further that "samples taken manually for settleability analyses typically contained SS concentrations much lower than concentrations of samples" taken by the slice-sampler. Compared to a Manning model 6000 sequential sampler, the slice-sampler collected samples having 1.5–7 times the SS concentration. The nonflow weighted average SS mass removal and net removal was 55% and 37%, respectively, with a TF of 3.1.

A 12-ft (3.7 m) diameter swirl with a Q_d of 6.8 Mgal/d (26,000 m^3/d) and a hydraulic loading of 40 gal/ft^2/min (84 m^3/m^2/h) was demonstrated in Syracuse, New York (Drehwing et al. 1979). Samples were obtained by sampling devices with adequate intake velocities. SS mass removal and net removal for this project averaged 52% and 18%, respectively, with a TF of 1.53.

An evaluation for the 17.3-ft (5.3 m) diameter Storm King vortex separator was undertaken at James Bridge in England (Hedges et al. 1992). Flows were pumped with controlled input materials. As the hydraulic loading doubled from 7.4–15 gal/ft^2/min (18–37 m^3/m^2/h), capture of wooden blocks remained perfect, but condom efficiency dropped off from 63%–39%. Testing of a 6-foot (1.9 m) diameter Storm King separator at the Wards Island WWTP in New York City showed that adequate capture of typical CSO floatables could be accomplished only at hydraulic loading of 10 gal/ft^2/min (25 m^3/m^2/h) where SS removal was 17%. At higher applied loadings, floatables capture deteriorated (URS 1990).

At an evaluation of the Lincoln Park vortex separator facility in Decatur, Illinois, which consists of four 45-ft (13.6 m) diameter Fluidseps and can handle a

Q_d of 400 Mgal/d (1,500,000 m^3/d), no visible floatables were reported escaping at an effluent hydraulic loading of 28.0 gal/ft^2/min (68.5 m^3/m^2/h) (Muhs 1995). For ten monitored storms the net SS removal (underflow volume effect excluded) averaged 39.1% with an average effluent hydraulic loading of 5.2 gal/ft^2/min (12.7 m^3/m^2/h). The facility also consists of two mechanical screening facilities: a 2 Mgal (7500 m^3) open, circular, first-flush storage tank, a 36-ft (11 m) diameter vortex flow dividing chamber where two asymmetric flow inputs of 250 and 164 Mgal/d (950,000 and 620,000 m^3/d) divided into the four 100 Mgal/d (390,000 m^3/d) waste streams to the four Fluidsep vortex solids separators.

A full-scale 200 Mgal/d (760,000 m^3/d) vortex solids separator demonstration facility is undergoing a year-long evaluation in New York City. This facility consists of three parallel vortex solids separators (swirl, Fluidsep, and Storm King) with each having a diameter of 43 feet (13.1 m). The purpose of the project is to fully document floatables- and settleable-solids-capture effectiveness for the three technologies under varying hydraulic loadings. Figure 4–14 gives the aspect (height to diameter) ratios of the three vortex devices.

Helical Bend Regulator/Concentrator

The EPA helical bend CSO flow regulator/settleable-solids concentrator (hereafter referred to as helical bend) is based on the concept of using the helical motion imparted to fluids at bends when a total angle of approximately 60 degrees and a radius of curvature equal to 16 times the inlet pipe diameter (D) are employed.

Figure 4–15 illustrates the device. The basic structural features of the helical bend are the transition section from the inlet to the expanded straight section be-

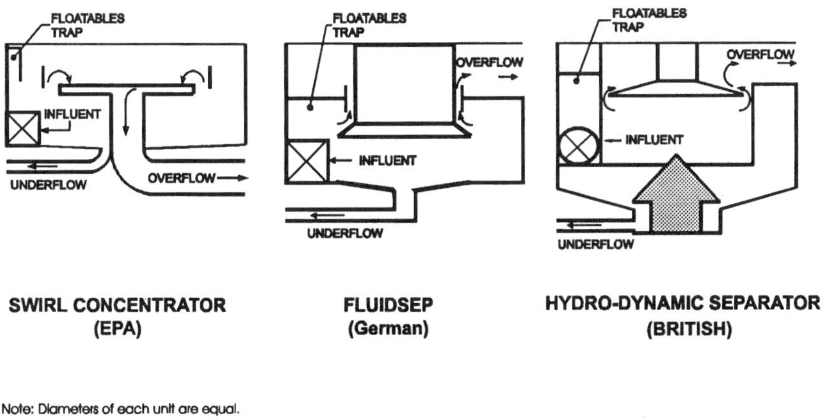

Figure 4–14. Vortex-treatment devices comparison.

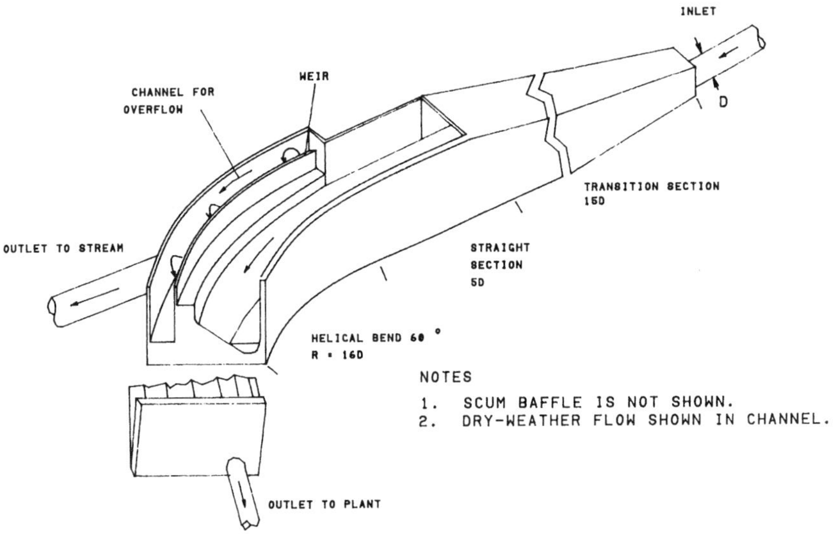

Figure 4–15. Helical-bend regulator/concentrator (Sullivan et al. 1982).

fore the bend, the overflow side weir and scum baffle, and the foul outlet for concentrated solids removal and control of the amount of underflow going to the treatment works.

DWF goes through the lower portion of the device and to the intercepting sewer. As the liquid level increases during wet weather, helical motion begins and the solids are drawn to the inner wall and drop to the lower level of the channel leading to the treatment plant. The relatively clean CSO passes over a side weir and is discharged to the receiving-water or storage and/or treatment facilities. Prevented from overflowing by a scum baffle along the side weir, floatables collect at the end of the chamber. They are conveyed to the treatment plant when the storm flow and liquid level subside.

Based on laboratory tests, pollutant removals in a helical bend are comparable to those in a swirl. Helical bends and swirls are, in effect, upstream treatment devices for the removal of relatively heavy, coarse material.

A comparison of construction costs for helical bend and swirl regulator/concentrators is presented in Table 4–6. These costs do not reflect the real cost-effectiveness of swirls and helical bends since these units actually serve dual functions: flow control and wastewater treatment. The construction cost for the helical bend is higher than for the swirl, but the helical may be more appropriate for a particular site, based on space availability and elevation difference between the interceptor and the incoming combined sewer (the helical bend requires a smaller elevation difference than the swirl).

Table 4–6. Comparison of Construction Costs for Helical Bend and Swirl Regulator/Concentrators (ENR 5 5,000), Land costs not included.

Capacity	Swirl	Helical
1.42 m^3/s (50 cfs)	$ 459,360	$1,121,500
2.83 m^3/s (100 cfs)	744,910	2,102,300
4.67 m^3/s (165 cfs)	1,034,595	2,963,090

Storage

Storage is considered a necessary control alternative because of the high volume and variability associated with CSO. Storage is also the best documented abatement measure currently practiced. Storage facilities are frequently used to attenuate peak flows associated with CSO. Past project results and theory indicate that storage must be considered at all times in system planning because it allows for maximum use of existing dry-weather treatment plant facilities and results in the lowest cost system in terms of treatment. The captured CSO is released to the WWTP when it can accept the extra volume during low-flow, dry-weather periods.

In comparison to treatment plants, storage facilities can respond without difficulty to intermittent and random storm behavior, are not upset by water quality changes, and are simpler in structural design. Disadvantages are large size, land requirements, and operation and maintenance (O&M).

Storage tanks' O&M problems stem mainly from floatables and solids deposition. Provisions in contemporary designs incorporate cost-effective means of dealing with floatables and deposited solids. Deposited solids can be removed after the event (as discussed later under "Storage Tank Cleaning by Flushing"). Floatables and deposited solids can be directed away from storage toward downstream treatment by using vortex separators (as previously pointed out in this chapter under "Vortex Separators"). Vortex separator devices also inherently provide a certain degree of storage. Chapter 5 illustrates the complimentary storage feature given a peak design flow of a one-year storm for Onondaga County.

CSO storage facilities can serve more than one purpose and include sedimentation treatment for excessive CSOs, DWF equalization (by compartmentalizing the storage basin for DWF capture and thereby reducing sludge handling problems) to enhance downstream DWF treatment and sewerline relief. These alternative functions should be used whenever possible.

Storage facilities can be classified as either *inline* or *offline.* Inline facilities are connected in series to the main combined sewer and generally create storage by letting the inlet orifice capacity surpass the discharge capacity. Offline storage is connected in parallel to the main sewer, and wastewater bypasses the storage facility during DWF or low-flow conditions. Inline storage usually does not require pumping. Inline storage can consist of either storage within the sewer pipes ("in-pipe" or "insewer") or in storage basins. Offline storage requires detention facilities (basins or tunnels) and usually requires facilities for pumping combined wastewater to storage or pumping it back to the sewer system.

Examination of storage options should begin with insewer storage. If this is not suitable, the use of inline storage tanks should be considered; however, sufficient head must be available as no pumps will be used. Offline storage should be considered last, since this will require power for pumping. Because the idea of storage is to lower the cost of the total treatment system, the storage capacity must be evaluated simultaneously with downstream treatment capacity so that the least cost combination for meeting water/CSO quality goals can be implemented.

Capital and O&M costs for basin storage facilities are presented in Table 4–7, and construction-cost curves are shown in Figure 4–16. Since the curves do not include pumping facilities, they are applicable to inline basins. The costs for earthen basins include liners.

Treatment by sedimentation should be taken into account for the design of these smaller, more economical, storage tanks. During periods when the tank is filled to capacity, the excess that overflows to the receiving water will have had a degree of primary treatment by way of sedimentation.

Storage-tank and basin design should incorporate the hydrology of the entire drainage system and the withdrawal rate from the WWTP and/or satellite treatment facilities. The design should be based on the historic rainfall records, using a continuous modeling approach to simulate long-term effects and dual function of sedimentation treatment (as previously discussed). A reliable answer to the required storage volume can be obtained only through the use of continuous modeling, as it does not assume that the storage tank is empty for every storm event. Figure 4–17 depicts this type of "mass-flow" analysis. The analysis by Driscoll et al. (1986) also incorporates the dependence on the storm interval.

Compartmentalization of the storage tanks can optimize their use in the collection system in these ways:

- A compartment of the storage tank can be set up, whether through an automated system measuring turbidity or historical statistical data, to accept the portion of the CSO event with the highest SS concentration. This will reduce effluent SS loadings to the receiving water and will reduce maintenance and sludge removal difficulties by (1) allowing relatively cleaner storm flow to fill the rest of the storage tank surrounding the compartment or by (2) utilizing compartments of storage tanks set up to overflow in series with the last compartment containing CSO with the lowest SS concentration configured to overflow to the receiving water.

Table 4–7. Summary of Basin Storage Costs[a] (Lager et al. 1977b)

Location	Storage capacity Mgal	Drainage area in acres	Capital cost in $	Storage cost in $/gal	Cost per acre in $/acre	Annual operation & maint. cost in $/yr.
Akron, Ohio (1)	1.1	188.5	1,139,195	1.03	6,054	7,242
Milwaukee, Wisconsin (Humboldt Ave.)	3.9	570	4,435,110	1.13	7,778	127,460
Boston, Massachusetts, Cottage Farm detection and chlorination,[b] station	1.3	15,600	16,236,960	12.50	1,043	202,780
Charles River marginal conduit project (2)	1.2	3,000	23,725,380	19.77	7,960	246,230
Chippewa Falls, Wisconsin	2.82	90	1,868,480	0.65	20,713	6,808
Sandusky, Ohio (3)	0.36	14.86	1,303,590	3.61	87,485	15,498
Washington, D.C. (4)	0.20	30.0	2,172,650	11.02	73,580	8,400
Columbus, Ohio (5) (Whittier Street)	3.75	29,250	15,353,420	4.10	521	—
Cambridge, Maryland (6)	0.25	20	796,640	3.20	40,560	36,210

[a]Estimated values.

[b]Estimated area.

(1) media-void space storage

(3) underwater

(4) underwater

(6) underwater

$/acre × 2.47 = $/ha

$/gal × 0.264 = $/l

Mgal × 3785 = m^3

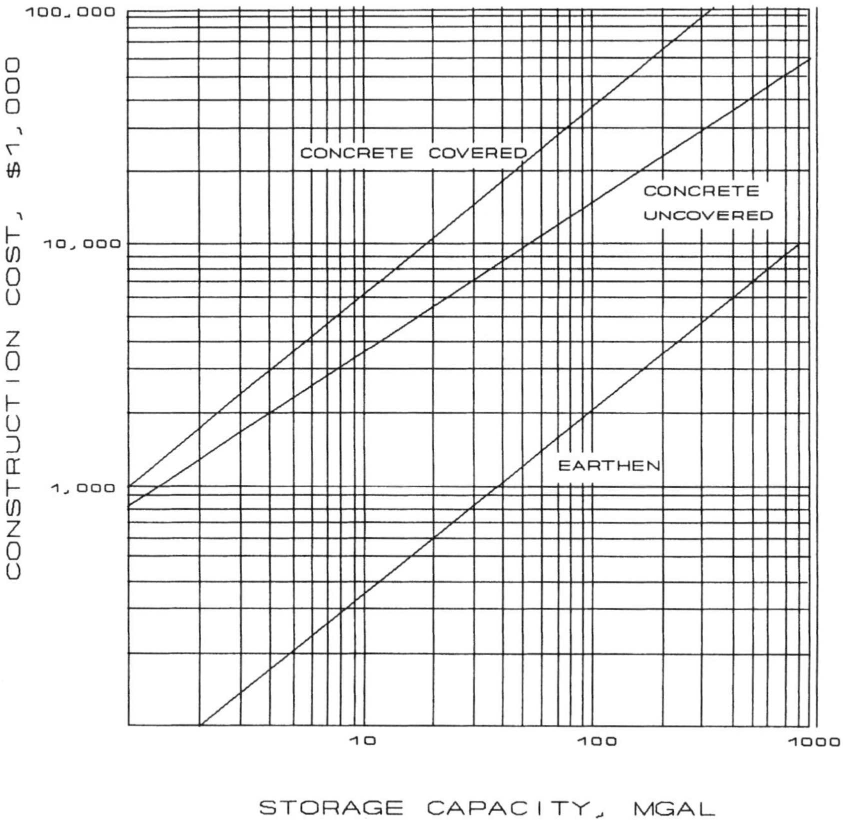

Figure 4–16. Storage-basin-construction costs (Lager et al. 1977b).

▪ Compartments can also equalize the diurnal variations in the relatively concentrated DWFs to the WWTP. The compartments could be used to capture the peak of the DWF to be released to the WWTP during low-flow periods, thereby creating a relatively steady-state loading to the WWTP. This dual-purpose operation of CSO storage tanks will optimize the use of the WWF facility, resulting in reduction or elimination of WWTP overloads, and afford easier DWF steady-state operation. The City of Rohnert Park, California (Welborn 1974), as part of a demonstration of the dual use of storage tanks for equalizing peak DWF as well as WWF, illustrated the economic advantages of their use as part of a regional wastewater handling scheme.

Even without treatment, flow attenuation helps equalize the pollutant loading to be assimilated by the receiving water and reduces the peak flows and consequent erosion in the receiving stream.

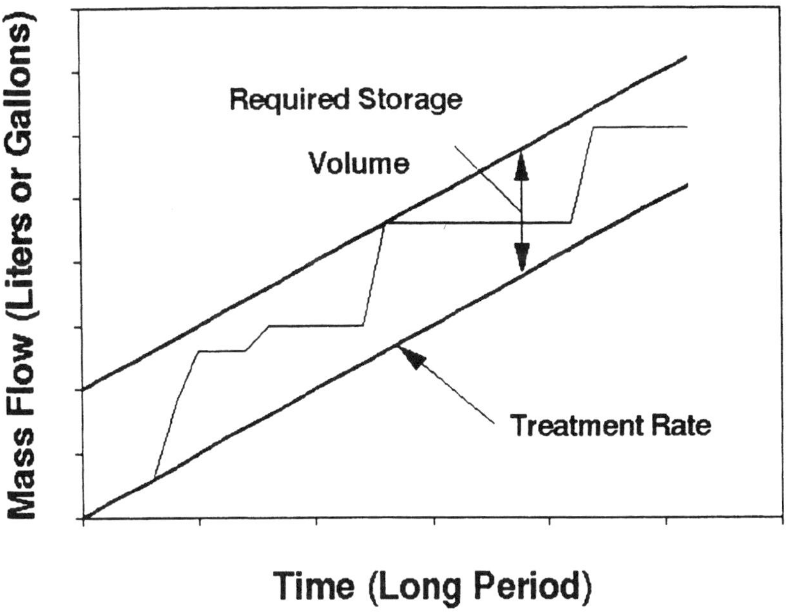

Figure 4-17. Long-term mass curve (Field and O'Connor 1997).

Inline Storage

Inline storage is the use of the available volume in the drainage system network of pipes and channels to store storm-generated flow. Inline storage capacity can also be provided by storage tanks, basins, tunnels, or surface ponds that are connected inline to the conveyance network. For maximum benefit, inline storage should be combined with some form of treatment; otherwise only flow attenuation will be achieved. Inline storage is unlikely to offer any treatment in itself through settling as the intent will be to make the system self-cleaning to reduce maintenance requirements. However, if the storage is combined with an end-of-pipe treatment, the flow attenuation will help equalize the load to the treatment process, hence optimizing the size of the treatment plant and costs.

Because combined sewers are designed to carry maximum flows occurring perhaps once in five years (50–100 times the average DWF), during most storms, there will be considerable unused volume within the conduits. Insewer storage is provided by damming, gating, or otherwise restricting flow passage, causing wastewater to back up in the upstream lines. The usual location to create the back-up is at the CSO regulator or overflow point, but the restrictions can also be located upstream.

The degree to which the existing conveyance system can be used for storage will be a function of the pipe or channel sizes that will provide the storage volume, the pipe or channel gradient (relatively flat lines are likely to offer the most storage capacity without susceptibility to flooding of low areas), suitable locations for installation of control devices such as sluice gates and weirs, and the reliability of the installed control. It will be essential that accurate details of the existing system be collected from field surveys and as-built drawings. This will allow the storage capacity, numbers and locations of controls, and risk of upstream flooding to be assessed. This will also be invaluable in new drainage-system design where conveyance pipes and channels can be up-sized and hydraulic controls can be designed into the system for added system storage and routing.

Minneapolis, Minnesota (Metropolitan Sewer Board 1971), Detroit, Michigan (Watt et al. 1975), and Seattle, Washington (Leiser 1974) employed unused storage capacity within the existing sewerage system to reduce the frequency and volume of CSO. Table 4–8 shows capital and O&M costs for insewer storage.

Offline Storage

Offline storage facilities can be located at overflow points or near DWF treatment plants. Typical storage facilities include lined- and unlined-earthen basins and lagoons and covered- or uncovered-concrete tanks. Tunnels are also used

Table 4–8. Summary of Insewer Storage Costs (ENR = 5,000)

Location	Storage Capacity (Mgal)	Drainage Area (acre)	Capital Cost ($)	Storage Cost ($/gal)	Cost per Acre ($/acre)	Annual O&M Cost ($/yr)
Seattle, WA						
Control & Monitoring System	—	—	8,749,000	—	—	182,500
Automated Regulator Stations	—	—	9,748,000	—	—	550,000
Total	17.8	13,120	18,497,000	1.06	1,400	733,000
Minneapolis–St. Paul, MN	NA	64,000	7,532,000	—	120	—
Detroit, MI	140	89,600	7,025,000	.06	80	—

NA = Not available
\$/a X 2.47 = \$/ha
\$/gal X 0.264 = \$/l
Mgal X 3,785 = m^3
(EPA-600/8-77-014)

where land is not available. Storage is achieved by diverting flow from the drainage system when a certain flowrate is exceeded. The diverted water is stored until sufficient capacity is available downstream. Many of the regulators listed under inline storage can be used to divert the flow once the predetermined flowrate has been exceeded.

Flow-Balance Method of In-Receiving Water Storage

Karl Dunkers, a Swedish inventor, developed an in-receiving-water storage system for stormwater runoff. This in-receiving-water-storage system or flow-balance-method facility (FBM) uses a wooden pontoon and compartmentalized walls of flexible PVC, fiberglass, reinforced or other plastic material for stormwater and CSO storage in the lake instead of a conventional concrete tank or lined-pond-equalization system, which is relatively expensive and requires land. The FBM, illustrated in Figure 4–18, performs by the plug-flow principle. The bottom is the lake bottom itself. The curtains are anchored to the receiving-water bottom by concrete weights. The FBM is always filled, either with polluted storm flow or lake water. There are several installations in Sweden, two of which have been in use for over 17 years.

During a storm, the storm flow enters the FBM and pushes the lake water from one compartment to another in a plug-flow pattern. When not receiving runoff, the stormwater is pumped to treatment, and the flow is reversed in the FBM with lake water filling up the facility. Thus, the lake water is utilized as a flow-balance medium. These units are sized to yield an effective in-water volume equal to the storage required for the storm size selected for design. Costs have been estimated to be one-fifth to one-tenth the cost of conventional land-side storage without inclusion of real estate costs. The systems seem to withstand wave action up to three feet (0.9 m) as well as severe icing conditions. Punctured walls are easily patched. Maintenance has proven to be inexpensive.

The Fresh Creek FBM is the first installation for either seawater or CSO and utilizes an operating principle based on density difference for displacement rather than plug flow. Fresh Creek is an inlet of Jamaica Bay in Brooklyn, New York City, and receives large CSO that degrade creek water quality. The FBM was constructed at Fresh Creek for demonstration, evaluation, and partial control of the CSO. Using only one bay, the creek water is displaced vertically by lower density CSO influent that forms a layer above the denser creek water.

There are two phases to the Fresh Creek demonstration project. Phase 1 (12/87 to 11/90) concentrated on proving the feasibility for CSO to displace creek water, form a stable layer, and be pumped back to the WWTP. It also examined the ability of the FBM to structurally withstand a marine environment including tidal exchange, freezing, and coastal storms. Phase 1 did not include any monitoring of water-quality variables. Phase 2 expanded the system capacity (at mean-tide level) from 0.41 Mgal (1550 m^3) to 2 Mgal (7570 m^3). The phase 1

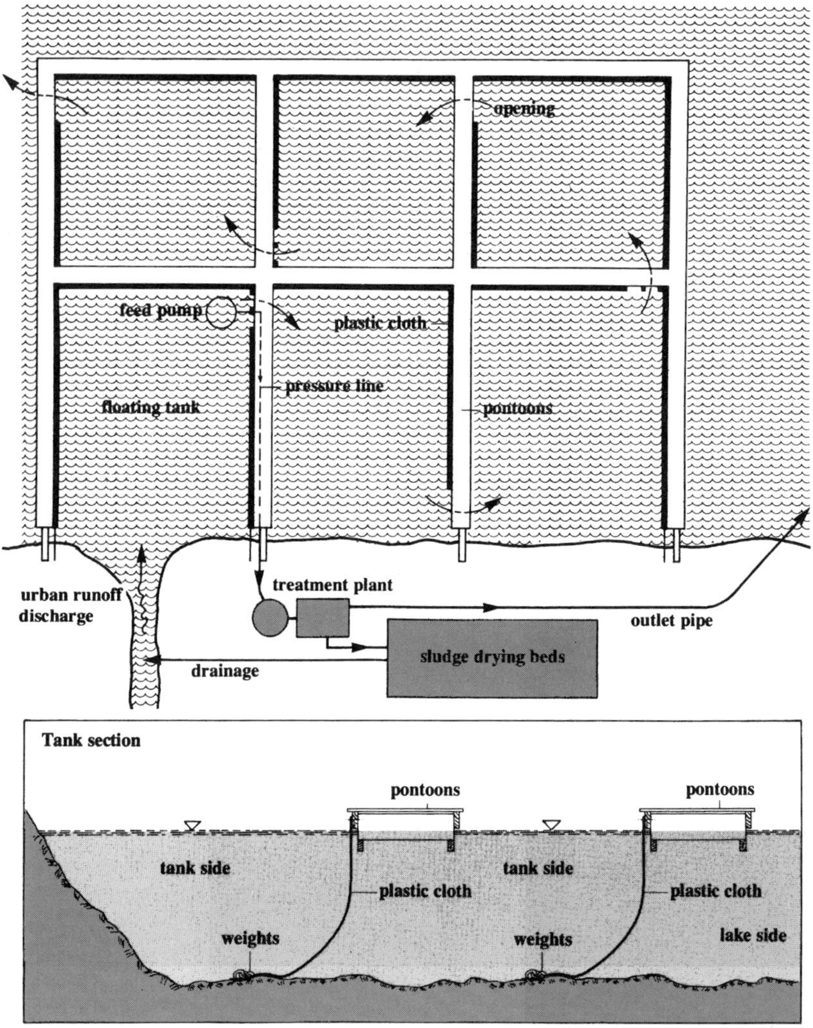

Figure 4–18. Flow-balance method facility (Karl Dunkers; Stockholm, Sweden).

study estimated hydraulic efficiency using specific conductivity as a tracer. Results from phase 2 are forthcoming.

Figure 4–19 is a plan view of the phase 1 installation. Four CSO outfall barrels form one side, the shore another, and the remaining sides are constructed from the previously described pontoon-and-curtain arrangement. Creek water moves into and out of the FBM through the end curtain farthest from the CSO barrels. This end curtain extends only part way to the creek bed, leaving an open-

Figure 4–18. Continued.

Figure 4–19. Fresh Creek installation (plan view), phase 1.

ing between the bottom edge of the curtain and the creek bed. Tidal exchanges are accommodated by allowing sufficient slack in the flexible curtain material. The plan layout of the facility is maintained by a system of wire rope ties and hollow tubular steel struts as shown in Figure 4–20.

Phase 1 successfully demonstrated principles of operation and the ability of the system to withstand the marine environment. The FBM showed no damage due to stresses caused by saltwater, tidal exchanges, CSO events, or coastal storms. The FBM was exposed to tidal ranges up to 7.0 feet (2.1 m), wind gusts up to 40 mi/h (64 km/h), and icy conditions. Stable stratification within the FBM was confirmed by taking several vertical specific-conductivity profiles during the FBM operation (Forndran et al. 1991; Field et al. 1990).

A ratio of pumpback to influent CSO volumes was also measured to evaluate hydraulic efficiency. However, because of down time and equipment problems, only three full sets of data were obtained. Table 4–9 records the influent, stored, and pumpback CSO volumes. Generally, the stored CSO volume is calculated by measuring the depth of the vertical specific-conductivity profiles and included the volume of CSO stored in the CSO barrels.

Hydraulic efficiency for the FBM of the first two events in Table 4–9 is only 3.2% and 5.0%, respectively (Field et al. 1995). Here the ratio of the pumpback volume to total influent CSO volume is very low because the FBM was a pilot and, accordingly, not adequately sized (due to financial restraints) to capture sig-

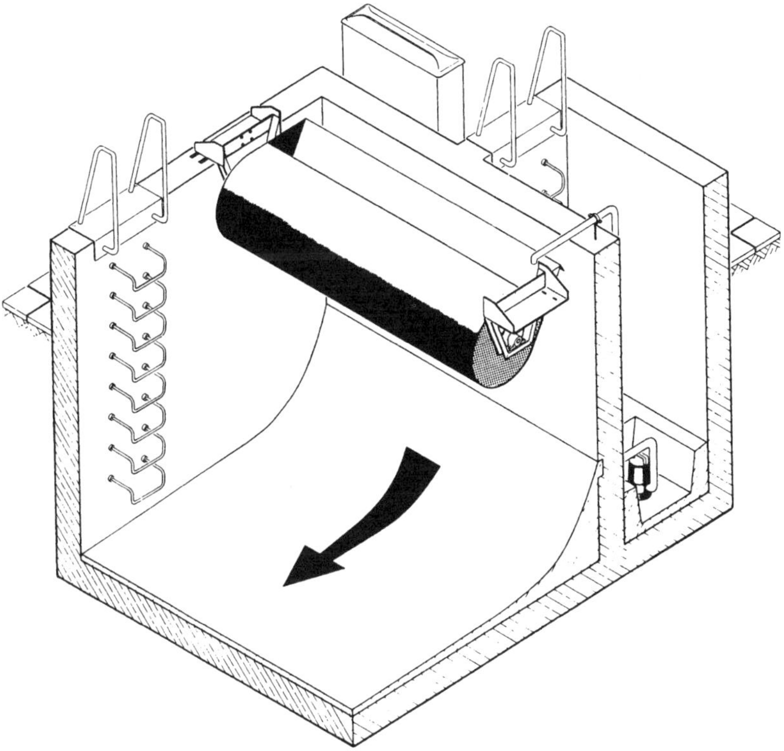

Figure 4-20. Sediment flushing tank.

nificant amounts of the CSO event volumes. The third CSO event showed the best overall efficiency (77%) as it was small and did not exceed the capacity of the FBM. This event also indicates that 23% of the CSO was lost from the FBM by diffusion and mixing across the boundary of the CSO and seawater layers (Field et al. 1995).

The FBM proved to be an efficient capture system for floatable material in the CSO influent. A system of net bags using the passive force of the flowing

Table 4-9. FBM Data for CSO Events

	Volumes (m^3)		Efficiency (%)
Event 1989	Influent CSO	Pumpback CSO	Total CSO Volume
7/16–17	60560	1968	3.2
9/14–15	25739	1287	5.0
9/17	984	757	77.0

CSO to concentrate the floatables in the bags and mechanical removal of the net bags has been incorporated for evaluation in phase 2.

The relatively low cost of the materials and construction gives this system advantages over conventional reinforced concrete tank systems (estimated to be 40–50% of conventional storage for saltwater applications). Total capital costs were $565,000 for a mean tidal volume of 0.41 Mgal (1550 m^3) of storage space in 1989, not including the previously mentioned savings from a significant reduction in required land area. However, any application of the FBM within a natural water body is subject to complying with federal, state, and/or local regulations.

Storage Tank Cleaning by Flushing

Two new methods for cleaning accumulated sludge and debris in CSO and stormwater storage tanks have emerged from Germany and Switzerland. Both methods create high-speed flushing waves to resuspend sediments on the tank floor and sweep these materials to a disposal channel at the end of the tank. The Hydroself (described under "Sewer flushing") uses flushing gates. This system has been used extensively in Europe, with 209 installations and 436 units in operation since 1985. Approximately 37% of the projects are to flush sewers, interceptors, and tunnels. The remaining installations are for cleaning sediments from CSO tanks. The first flushing-gate system installed in North America is for a 3.2-Mgal (12,000 m^3) CSO tank in Sarnia, Canada (Parente et al. 1995). The second method uses *tipping flushers,* which were initially developed in Switzerland. In Germany, independent research optimized this cleaning system to its present efficient design (Brombach 1987a). In Europe, there are more tipping-flusher installations than flushing-gate systems.

The tipping flusher is a cylindrical stainless steel vessel suspended above the maximum water level and affixed to the backwall of the storage tank (Figure 4–20). After the tank is drained following an event, a local water supply (e.g., river water, groundwater, potable water, or sewage) is pumped into the tipping flusher at low rates. Just prior to the tipping flusher overflowing with the filling water, the center of gravity shifts, and the tipping flusher rotates about its central axis, thereby spilling its contents along the backwall of the tank. The flushing wave spills uniformly on the wall and accelerates. An arch at the intersection of the backwall and tank floor directs the flushwater across the floor of the tank without substantial energy loss. The flushwater removes the sediments from the tank floor and transports it to the sump located at the far end of the tank.

The required water volume depends on the:

- flushing distance
- height of the tipping flusher above the tank floor
- tank floor slope in the direction of flushing.

The tipping flusher volumes range from 12–120 gal/ft (150–1500 l/m) with maximum flushing lengths of 33–150 feet (10–50 m). At 150 feet (50 m), the energy of the flushing wave cannot be sustained, dissipating into rivulets at greater flushing lengths. Two flushes are often necessary in the 160–260-ft (50–80 m) range.

These systems require flushing lanes nearly equal to the width of the tipping flushers (leaving enough space for flusher support). These lanes require a training wall 1.3 feet (0.4 m) high along the tank length to the discharge channel. The tank side walls are square with the floor to ensure the wall edges are cleaned. The back wall (with flushers attached) is typically arched at the bottom. The upper tipping-flusher effective height limit is about 18 feet (5.5 m). The tipping flusher should be situated above the maximum expected water level in the tank to avoid its submergence.

Tipping flusher systems have been in use in North America for five years (about 25 systems in the United States, with most in the state of Michigan). These systems have been employed in Germany and Switzerland since about 1987. Some installations are completely automated; none of the European installations have required any replacement of parts or additional maintenance (due to the absence of moving parts except for the bearings located at each end). In Germany and Switzerland, tank side walls are generally hand troweled to a very smooth finish to alleviate secondary wall cleaning. If there are no protrusions (columns, mixers, etc.) in the flushing lane and if the flushing lengths are <150 feet (<50 m), little secondary cleaning of the tank floor is required. Similarly, with adequate slope and rapid drain capacity, little secondary cleaning is necessary. End trough cleaning, if necessary, is usually done by hosing.

As with inline storage, offline storage can be designed to be relatively self-cleaning or have facilities to resuspend the settleable solids.

TREATMENT

Physical/Chemical

Physical processes with and without chemical additives are of particular importance in CSO treatment because of their adaptability to automatic operation (including almost instantaneous startup and shutdown), excellent resistance to shockloads, and ability to consistently produce a low SS effluent.

Screening

Screens have been used to achieve various levels of SS removal and may be used for three process applications:

- Main treatment—Screening is the primary treatment process.
- Pretreatment—Screening removes suspended and coarse solids prior to further treatment to enhance the treatment process or protect downstream equipment.
- Dual use—Screening provides either main treatment or pretreatment of storm flow and is used as either an effluent polisher or for main treatment during periods of DWF.

Screens can be divided into four categories (with apertures):

1. bar screens (>1 inch [>25.4 mm])
2. coarse screens (1–3/16 inch [25.4–4.8 mm])
3. fine screens (3/16–1/250 inch [4.8–0.1 mm])
4. microscreens (<1/250 inch [<0.1 mm]).

Solids removal efficiencies for screens are affected by two mechanisms:

- straining by the screen, and
- filtering of smaller particles by the mat (*schmutzdecke*) deposited by the initial straining.

The efficacy of these mechanisms is, in turn, a function of the screen aperture and SS characteristics (concentration and particle-size distribution). Typically the efficiencies of screens treating a waste stream with a given distribution of particle sizes will increase as the screen aperture decreases.

The first condition makes these devices very versatile as the removal efficiency of screening devices is adjustable by changing the aperture size of the screen placed on the unit. The second condition affecting removal efficiencies is the thickness of filtered material on the screen. Whenever the thickness of this filter mat is increased, the suspended matter removal will also increase because of the decrease in effective pore size and the filtering action of the filtered mat. However, a thicker mat will also cause a larger head loss across the screen and is a limiting condition in design.

The basis for bar and coarse-screen design is the same as for their uses in DWF treatment facilities. Because CSO contains a significant amount of coarse debris, which is aesthetically undesirable, providing coarse screening as the minimum CSO treatment may be useful.

Microscreening (Lager et al. 1977) was effectively demonstrated for CSO primary treatment in Philadelphia, Pennsylvania (Maher 1974), and Syracuse, New York (Drehwing et al. 1979). In Philadelphia, a 23-mm aperture Microstrainer attained average SS removals of 70% with increased SS removals of 78% using polyelectrolyte addition (0.25–1.5 mg/l). The hydraulic loading was 16 gal/ft^2/min (39 m^3/m^2/h) without and 36 gal/ft^2/min (88 m^3/m^2/h) with polyelectrolyte addition. A laboratory coagulation study found moderately charged, high-molecular-weight cationic polyelectrolytes to be the most suitable for this application. The influent concentration and removal efficiency would vary, whereas effluent concentration remained relatively constant. This phenomenom is apt to recur in other physical-chemical stormwater treatment operations (Field and Struzeski 1972).

The Department of Drainage and Sanitation, Onondaga County (Syracuse), New York, demonstrated Zurn 71-mm and a Crane 23-mm aperture microscreening units. Average SS removals were 48% at hydraulic loadings between 5 and 62 gal/ft^2/min (12 and 150 m^3/m^2/h) and 58% at hydraulic loadings between 3.3 and 13.7 gal/ft^2/min (8.1 and 33.5 m^3/m^2/h), respectively.

In Deerfield, Illinois (Gavle and Mitchel 1996), Hycor horizontal rotating screens were installed as part of two SSO satellite-treatment facilities that have been in operation for over five years. SSO is relatively similar to CSO in that it is storm generated and has a sanitary wastewater constituent diluted by stormwater. The screen apertures were 0.04 inch (1 mm) (fine-screen range). The SSO is discharged inside the screen and collected in a large trough below the screen. The facilities are capable of maintaining concentrations of BOD_5 below 44 mgl and SS below 49 mg/l. One facility had higher SS effluent concentrations in the first two years of operation.

The continuous deflective separation (CDS) device (CDS Technologies) was originally designed for removal of gross pollutants from stormwater. It consists of a cylindrical chamber containing a circular, perforated (screenlike) plate through which the stormwater passes. The cylindrical shape creates a continuous circular flow pattern over the screen and prevents accumulation of debris that occurs in direct-screening processes (Wong and Wooton 1995). This flow pattern also decreases the effective aperture and increases performance; actual screen apertures for these devices range from the coarse to the fine screen size. CDS is currently being demonstrated in Australia. Figure 4–21 shows a plan and elevation drawing of a CDS device.

Dissolved-Air Floatation

Dissolved-air floatation (DAF) separates solid particles or liquid droplets from the liquid phase by introducing fine-air bubbles into the liquid phase. As the bubbles attach to the solid particles or liquid droplets, the buoyant force of the particle/droplet and air bubbles is great enough to cause the particle to rise. The particles float to the surface and are removed by skimming. Air bubbles are usually first dissolved into the wastewater stream under pressure. The dissolved air then comes out of solution as minute bubbles. The pressurized flow carrying the dissolved air to the floatation tank can be comprised of (1) the entire storm flow, (2) a portion of the storm flow (split-flow pressurization), or (3) recycled-DAF effluent.

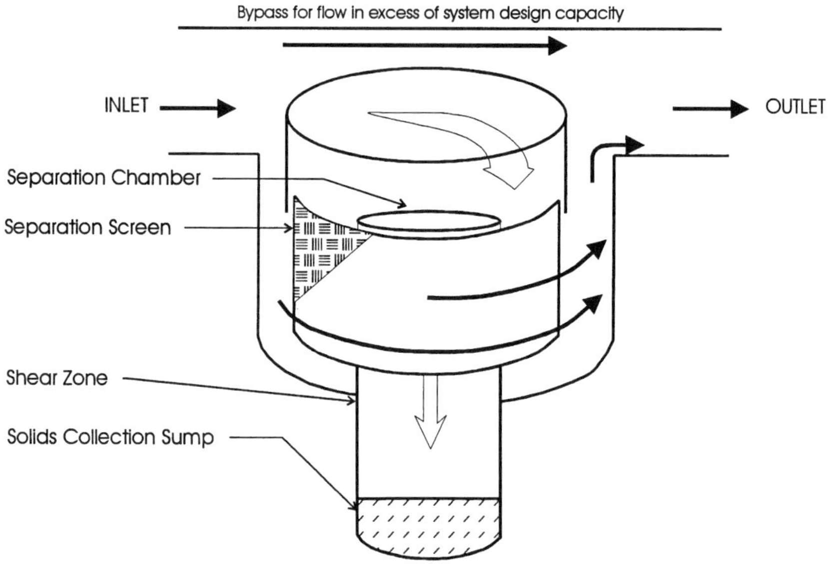

Figure 4–21. Continuous deflective separator (CDS).

Higher hydraulic loadings (1.3–10.0 gal/ft^2/min [3.2–25 m^3/m^2/h]) and shorter detention times (0.2–1.0 hour) can be used for DAF than for conventional settling (0.2–0.7 gal/ft^2/min [1.7 m^3/m^2/h]; 1.0–3.0 h). This process has a definite advantage over gravity sedimentation when used on CSO, since particles with densities both higher and lower than the liquid can be removed in one skimming operation. DAF also aids in the removal of oil and grease, which are not as readily removed during sedimentation.

The principal parameters that affect removal efficiencies are:

- hydraulic loading or overflow rate,
- amount of air dissolved in the flows, and
- chemical additives (ferric chloride is the most common).

Typical design parameters for DAF facilities are presented in Table 4–10.

Screening followed by DAF effectively reduced pollutants in CSO during a demonstration project (Gupta et al. 1977). The screening (aperture 297 μm) removed the larger particles too heavy for the air bubbles to float. Average reported percent removals at Milwaukee (5 Mgal/d [18,925 m^3/d] pilot), with and without chemical addition, are listed in Table 4–11. The chemical flocculants required to achieve these removals were 20 mg/l ferric chloride and 4 mg/l cationic polyelectrolyte.

Table 4–10. DAF Design Parameters

Overflow Rate (gal/ft²/min)	
Low rate	1.3–4.0
High rate	4.0–10.0
Horizontal Velocity (ft/min)	1.3–3.8
Detention Time (min)	
Floatation cell range	10–60
Floatation cell average	25
Saturation tank	1–3
Mixing chamber	1
Pressurized Flow (% of total flow)	
Split-flow pressurization	20–30
Effluent recycle pressurization	25–45
Air-to-Pressurized Flow Ratio (standard ft³/min/100 gal)	1.0
Air-to-Solids Ratio	0.05–0.35
Pressure in Saturation Tank (lb/in.²)	40–70
Float	
Volume (% of total flow)	0.75–1.4
Solids concentration (% dry weight basis)	1–2

gal/ft²/min X 2.445 = m³/m²/h
ft/min X 0.00508 = m/s
standard ft³/min/100 gal X 0.00747 = m³/min/100 l
lb/in.² X 0.0703 = kg/m²
(EPA-600/8-77-014)

At Racine, Wisconsin (Meinholz et al. 1979), two full-scale prototype plants (14.3 and 44.4 Mgal/d [54,126 m^3/d and 168,054 m^3/d], respectively) were demonstrated for the treatment of CSO using 40 mg/l ferric chloride and 2 mg/l cationic polyelectrolyte. The treatment efficiencies (percent removals on a concentration basis) are presented in Table 4–12. The results from Site II are better than for Site I because the hydraulic loading was usually lower at the former. Capital costs (including land) were $11,000,000 and $2,100,000 for Site I and II, respectively. The O&M costs for the systems were $0.58/1,000 gal ($0.16/m^3).

Carrier-Enhanced Settling

Microcarrier systems such as the Microsep (International Water Solutions Corporation) and the Actiflo (Omnium de Traitement et de Valorisation [OTV]) use recycled microcarriers (e.g., microsand [0.05–0.2 mm]), while DensaDeg

Table 4-11. Percent Removals Achieved with Screening/DAF (Milwaukee, WI) (Gupta et al. 1977)

	Without Chemical Flocculant Addition	With Chemical Flocculant Addition
BOD_5	35± 8	60±11
COD	41± 8	57±11
SS	43± 7	71± 9
Volatile SS	48±11	71± 9
Nitrogen	29±14	24± 9

(EPA-600-77-069a)

(Infilco Degremont) uses recycled sludge as the carrier to enhance settling. Each system adds chemicals to the recycled material to create heavier/larger floc that does not require as long a detention time for settling as do conventional settling tanks. The Actiflo claims 80% removals of SS with effluents of 30 mg/l, a 15-minute detention time, an ability to handle variable hydraulic loadings, and peak hydraulic loadings of 52 gal/ft^2/min (130 m^3/m^2/h) as compared to approximately 0.5 gal/ft^2/min (1 m^3/m^2/h) for conventional settling tanks. The Microsep claims 90% removal of SS, rapid startup (for offline installations), and a footprint that is 70% smaller than conventional settling tanks. The DensaDeg system claims hydraulic loadings from 10–14 gal/ft^2/min (17–24 m^3/m^2/h) and a 50% reduction in size as compared to conventional solids contact clarifiers while incorporating gravity sludge thickening.

Table 4-12. Percent Removals Achieved with Screening/DAF (Racine, WI) (Meinholz et al. 1979)

	Percent Removal	
	Site I	Site II
BOD_5	57.5	65.4
TOC	51.2	64.7
SS	62.2	73.3
Volatile SS	66.8	70.9
Total Phosphorus	49.3	70.0

(EPA-600/2-79-106a)

Biological Treatment

Biological treatment is a means for stabilizing dissolved organic matter and removing nonsettleable colloidal solids. This can be accomplished either aerobically or anaerobically. Several biological processes have been applied to CSO treatment. These include: contact stabilization, trickling filtration, rotating biological contactors (RBC) (shaft-mounted rotating disks), and treatment lagoons. Final clarification greatly influences the overall performance of these systems (optional for treatment lagoons) by preventing the carryover of biological solids produced by the processes.

Secondary biological treatment processes at WWTP are capable of removing between 70 and 95% of the BOD_5 and SS at dry-weather design flowrates and loadings. When these biological treatment processes are used for combined sewage treatment, removal efficiencies are lower due to higher hydraulic and lower organic loading rates. Also, most biological systems are susceptible to overloading conditions and shock loads as compared to physical treatment processes. Because of this limited ability of biological systems to handle fluctuating and high hydraulic shock loads, storage/detention facilities preceding the biological processes may also be required.

Secondary biological treatment options for retrofitting and new WWTP installations include: high-rate trickling filtration (deep honeycomb plastic media) utilizing two filters in parallel during WWF operation and automatically switching back to series during DWF operation (Homack et al. 1973); contact stabilization during WWF periods automatically switching to conventional activated sludge during DWF periods (Bennedict and Roelfs 1981; Agnew et al. 1975); RBC (Welsh and Stucky 1974); and biological-fluidized-bed filtration (Scheible and Grey 1988). These processes have been demonstrated to treat WWF at hydraulic loadings greater than DWF.

General maintenance problems experienced by wet-weather biological facilities are similar to those experienced at conventional biological installations. Winter operation of mechanical surface aerators have had some serious drawbacks, including icing, tipping, or sinking. Other promising methods of providing the required oxygen have been demonstrated at many dry-weather facilities and include diffused air systems and submerged tube aerators.

A comparison of construction and O&M costs for biological treatment systems and treatment lagoons is presented in Table 4–13. Costs of final clarification are included where control of solids and sludge produced by the biological treatment system are required. Costs also include pumping, disinfection, and algae control systems, when applicable. Engineering, administration, and land costs are not included in the estimates; however, land costs may be the controlling economic factor in the evaluation of lagoon treatment systems and therefore must be evaluated for each specific location. Biological CSO treatment

Table 4–13. Summary of Capital and Operation and Maintenance Costs for Biological Treatment Alternatives (ENR = 5,000) (Lager et al. 1977b)

Project Location	Type of Biological Treatment	Peak Plant Capacity (Mgal/d)	Construction Cost ($)	Cost/ Capacity ($/Mgal/d)	Cost/ Tributary area ($/acre)	Annual O&M ($/1,000 gal, except as noted)
Kenosha, WI	Contact stabilization	20	3,410,000	170,000	2,900	34.5
Milwaukee, WI*	Rotating biological contactor	4.3	747,000	173,000	21,400	11.0
Mount Clemens, MI						
Demonstration system	Aerated treatment lagoons	64	1,607,000	25,000	7,600	50.0
Citywide system	Storage/aerated treatment lagoons	260	14,343,000	55,000	9,800	47.5
New Providence, NJ†	High-rate trickling filter	6	1,188,000	198,000	—	30.7
Shelbyville, IL						
Southeast site	Oxidation lagoon	28	108,000	3,900	2,500	3,820/yr‡
Southwest site	Storage & facultative lagoons	110	844,000	7,700	1,900	14,460/yr‡
Springfield, IL	Oxidation lagoon	67	440,000	6,500	200	5,260/yr

*Includes estimate of final clarifier.
†Includes plastic media trickling filter, final clarifier, plus one-half of other costs.
‡Based on estimated man-day labor requirements.

Mgal/d X 0.0438 = m^3/s
acre X 0.405 = ha
¢/1,000 gal X 0.264 = ¢/m^3

(EPA-600/8-77-014)

systems generally are integrated with or are a part of dry-weather treatment facilities. Estimates of the wet-weather costs of these facilities were separated from total costs of the total treatment systems. The cost of the inline RBC at Milwaukee, WI, was used together with an estimated cost for a final clarifier to develop an estimated cost of a complete RBC treatment system. The final clarifier cost was based on one 65-ft-(19.8 m) diameter clarifier with a surface loading rate of 0.83 gal/ft^2/min (2.0 m^3/m^2/h). Costs of lagoon treatment systems vary widely and are a function of the type of lagoon (oxidation, aerated, or facultative), the number of cells, and the miscellaneous equipment requirements including aeration, disinfection, instrumentation, pumping, and algae control. Costs for many of these CSO facilities are based on only one installation of each biological treatment process. Therefore, these costs should be considered only rough estimates and may be greatly influenced by the degree of integration with DWF treatment required to produce a viable system. These costs can be used as a preliminary guide, but detailed analyses should be performed to compare and evaluate biological treatment alternatives with other methods of treatment and control.

Initial capital investments of integrated dual-use facilities can be reduced by apportioning part of the costs to the DWF facility. The cost reduction is in proportion to the net benefit that the wet-weather facility provides to the overall treatment efficiency during DWF periods.

Trickling filters. A high-rate trickling filter was constructed at the New Providence, New Jersey, WWTP to alleviate hydraulic overloading and resultant loss of treatment efficiency caused by excessive sanitary sewer infiltration (Homack et al. 1973). The plant utilized two high-rate trickling filters, one with conventional rock media and the other with deep-bed-honeycomb-plastic media operating in parallel to treat WWF. During DWF periods the plant operated in series with a controlled flow to maintain an active biological slime on the filters that also served to improve DWF treatment efficiency. The trickling filter operation automatically switched from series to parallel when the WWF exceeded 2.8 Mgal/d (11,000 m^3/d) and reverted to series when the flow fell below 2.0 Mgal/d (7,600 m^3/d). This type of operation provides greater capacity-enabling effective WWF treatment.

This investigation proved it is technically and economically feasible to construct and operate a treatment plant using a combination of series-parallel high-rate trickling filters. It was determined that the hydraulic loading should be limited to 70 million gallons per acre per day (Mgal/acre/d) (2.7 m^3/m^2/h) and the organic loading rate to 85 lb BOD_5/1000 ft^3 (0.41 kg/m^2/d) on the plastic-media filter and 20 Mgal/acre/d (0.78 m^3/m^2/h) and 40 lb BOD_5/1000 ft^3 (0.20 kg/m^2/d) on the rock-media filter to achieve an effluent containing 25 mg/l BOD_5 or less during periods of excessive infiltration. The DWF BOD_5 and SS removal efficiencies were each approximately 93%. The corresponding concentrations of BOD_5 and SS in the final effluent are 9 mg/l 12 mg/l, respectively. WWF BOD_5 and SS removal efficiencies were each approximately 86%.

The plastic-media-trickling filter removed about 2.7 times more BOD_5 than the rock media. Accordingly, it was recommended that the plant operate at 2:1 flow split between the plastic/rock media filters during WWF. Although initial cost of the plastic media filter was 25% more than the rock, the overall ratio of cost per pound of BOD_5 removed per 1000 ft^3 was 2:1 in favor of the plastic media type.

A drawback of the high-rate trickling filter process for the treatment of high flows is the tendency for colloidal suspended material to be discharged from the filters. This requires subsequent treatment (i.e., chemical precipitation and low secondary clarifier overflow rates or supplemental biological flocculation in the form of a pond) to achieve high SS and BOD_5 removals.

Contact Stabilization

In Clatskanie, Oregon, a full-scale, dual-use facility was constructed to alleviate SSO caused by excessive infiltration (Bennedict and Roelfs 1981). The plant treated WWF as high as 6.6 times DWF of 0.19 Mgal/d (720 m^3/d). The plant's primary physical and secondary biological treatment processes were both modified to convert automatically from conventional sedimentation and activated sludge treatment for DWF periods to DAF and contact stabilization for WWF periods since DAF and contact stabilization can accommodate higher hydraulic loadings. DWF BOD_5 and SS removal efficiencies were each 94%; WWF BOD_5 and SS removal efficiencies were each approximately 73% and 71%, respectively.

Contact stabilization sludge treatment relies on the sorptive ability of biological solids to remove biodegradable organic matter with short detention periods (Bennedict and Roelfs 1981). These solids are settled out in the secondary clarifier, returned to the aerobic digester for stabilization, and recycled to mix with the influent of the contact chamber. The recycle flow determines the contact chamber time and aerobic stabilization time. The contact, aeration, and sludge reaeration compartments are further subdivided into operating zones to provide flexibility. In general, aeration volume requirements of contact stabilization, a completely mixed system, are typically 50% less than conventional plug flow (Tchobanoglous and Burton 1991). The DAF-contact stabilization capital and total costs were only 14% and 10%, respectively, higher than the costs of a standard DWF plant. In Clatskanie it was estimated that dual treatment as an infiltration/inflow or SSO control method for increased WWF costs less than half as much as either sewer rehabilitation or flow equalization.

Treatment Lagoons

Pollutant removal efficiencies by treatment lagoons have varied from highs of 85–95% to negative values due to excessive algae production and carryover.

In addition to the type of lagoon and the number of cells in series (stages), several major factors that influence removal efficiencies include:

- detention time
- source of oxygen supply
- mixing
- organic and hydraulic loading rates and
- algae removal mechanisms.

A single cell storage/oxidation lagoon in Springfield, Illinois (Springfield Sanitation District 1970), provided an average BOD_5 removal of 27%, and SS effluent concentrations tended to range between 20–40 mg/l regardless of influent concentrations with the exception of higher effluent concentration during peak algae densities in the summer months. Notably, the facility reduced fish kills in the receiving water. At Mount Clemens, Michigan (Mahida and DeDecker 1975), construction of multiple lakelets resulted in annual removals of 94–96% of SS and 92–96% of BOD_5. Multiple-cell facilities with algae control systems constructed at Shelbyville, Illinois (Parks et al. 1974; Connick et al. 1981), provided 75–90% SS and BOD_5 removal efficiencies, respectively during WWF conditions.

Biological Aerated Filters

Biological aerated filters (BAF) use fine-grain media. This media has a dual function by retaining solids and biomass in the reactor and serving as a nucleus for biofilm attachment. Because the bacteria are fixed on the media, with aeration high concentrations of active biomass can be reached. The filter media grain sizes range from 2–10 mm. BAF combine unit processes to produce one compact reactor (i.e., aeration, biofilm layering, and filtration) thereby eliminating the need for downstream SS removal or clarification.

To date, more than 200 plants have been built, mostly in Japan, France, Germany, Britain, and Scandinavia. BAF is capable of handling flows from small industrial facilities to large municipal installations with flows of 100 Mgal/d. The compact size of the bioreactor allows it to be covered, amenable to odor control, and integrated into sensitive urban sites or environments. Different technologies can be chosen depending on treatment objective and operating parameters. Table 4–14 lists a few variations of BAF. Flow into the biofilter can be upflow or downflow, while air can be blown in at the base of the filter or in the middle of the bed. Most reactors also allow unaerated operation for dentrification. The media is most often expanded schist or clay, but recently lightweight plastic media has also been used. In most designs, media is retained by plastic nozzles in a floor or ceiling.

BAF was first introduced in North America in the early 1980s. However, the technology did not find acceptance at the time because of design and operational

Table 4-14. Biological Aerated Filters

Type	Company	Media	Water	Backwash
Biocarbone	OTV (F)	Shale	Down	Countercurrent
Biofor	Degremont (F)	Clay	Up	Cocurrent
Biostyr	OTV (F)	Polystyrene	Up	Countercurrent
Biodrof	Degremont (F)	Clay	Down	Countercurrent
SAFE	PWT (GB)	Shale	Down	Countercurrent
Colox	Tetra (US)	Clay	Up	Cocurrent
Biopur	Sulzer (CH)	Structured Plastic	Up	Cocurrent
BAF	Stereau (F)	Clay	Down	Countercurrent
Biofilter	CISE (F)	Pouzzolane	Up	Cocurrent
Biobead	Brightwater (GB)	Polypropylene	Up	Mechanical
Biofilter	Wabag (D)	Structure Plastic	Up	Cocurrent

problems and because compact reactors were not required. In the late 1980s, more than ten large plants using BAF were built in Quebec, Canada, where harsh-weather conditions favor compact installations. The largest BAF (peak flowrate > 20 Mgal/d) in Canada, which began operation in 1991, is in the municipality of Sherbrooke. Several 1-Mgal/d BAF will begin operation in 1997 in New York City for nitrogen removal. Two 5-Mgal/d BAF were installed in 1995 to polish and nitrify refinery effluent in Carson, California.

BAF require periodic backwashing to remove filtered solids and excess biomass. Headloss increases over time with maximum values of 6.9 feet (2.0 m) and intervals between backwashes varying from 24–48 hours under normal design conditions. The buildup of headloss in a downflow system tends to be lower than in an upflow system; however, downflow system's headloss increases exponentially at the end of the cycle, whereas upflow systems behave linearly.

The backwash hydraulic loading ranges from 8–50 $gal/ft^2/min$ (20–80 $m^3/m^2/h$); air scouring, as in other filter backwash operations, is also required. The sludge water generated from removing excess biomass and assimilated and trapped solids can be either thickened separately or returned to the primary settlers (adding about 10% of flow). Backwash typically lasts <0.5 hour and can be conducted in periods of low flow or low-energy cost.

BAF can be designed for carbon removal, nitrification, and/or dentrification. All three reactions can be carried out in the same reactor. Phosphorous precipitation has been combined with biofiltration on a large scale. Biological phosphorus removal has been developed using a pilot-scale facility. Although not universally suited for all situations, the flexibility of the BAF make them attractive tools for plant extensions and upgrading, including the biological treatment of CSO.

Disinfection

Conventional municipal sewage disinfection processes generally use chlorine (Cl_2) gas or sodium hypochlorite (NaOCl) due to their excellent oxidation capabilities and relatively low cost. Contact times of a minimum of 15 minutes at peak flow are commonly recommended.

In the last 20 years wastewater disinfection has come under scrutiny. Research studies, particularly for drinking water, have cited health risk with regard to Cl_2 and its byproducts. Excess free Cl_2 provides the basis for chlorinated hydrocarbon formation (chloroform and trihalomethanes [THMs]), which are suspected carcinogens. Cl_2 disinfection of wastewater usually results in excess free Cl_2 due to the variability of Cl_2 applied to satisfy the demand of ammonia (NH_3) and nitrate (NO^{-2}) compounds in wastewater. Research also indicates that Cl_2 residuals and compounds discharged to natural waters may be harmful to aquatic life.

Disinfection of CSO is generally practiced to control the discharge of pathogens and other microorganisms in receiving waters. However, the CSO have characteristics of intermittent, high flowrate, high SS content, wide temperature variation, and variable microorganism quality. Several other aspects of disinfection practices require consideration for CSO treatment applications:

- Coliform count is increased by surface runoff in quantities unrelated to pathogenic organism concentration. Total coliform or fecal coliform levels may not be the most useful indication of disinfection requirements and efficiencies.
- Discharge points requiring disinfection are often in outlying areas on the sewer system and require completely automated installations.
- Effective SS removal prior to disinfection will greatly enhance the disinfection process.

The disinfectant used at a facility for treatment of CSO should be adaptable to intermittent use. Other considerations include the disinfection effectiveness and the safety and ease of feeding. Table 4–15 shows disinfectants that might be used for CSO disinfection. Cl_2 and hypochlorite (OCl^-) will react with NH_3 to form chloramines and with phenols to form chlorophenols. These are toxic to aquatic life (and the latter also produce taste and odor in the water). Chlorine dioxide (ClO_2) does not react with NH_3 and completely oxidizes phenols. Ozone (O_3) is also effective in oxidizing phenols.

Another disinfectant that could be considered for future applications is bromine (Br). However, certain factors limit its application at this time. Although bromine can be shown to be a more effective disinfectant (Alleman 1988), it may pose a greater toxicity concern to the aquatic system (White 1986).

Table 4-15. Disinfection Technologies*

Disinfectant	Dose Range
Chlorine dioxide (ClO_2)	2–6 mg/l
Ozone (O_3)	25–80 mg/l
Sodium hypochlorite (NaOCl)	12–20 mg/l
Ultraviolet light (UV)	31–65 mJ/cm^2

*Disinfection technologies tested by NYC-DEP as part of the Spring Creek Disinfection Study

High-rate disinfection (i.e., decreased disinfection contact time) can be accomplished by:

- increasing mixing intensity
- increasing disinfectant concentration
- using faster-acting oxidants and ultraviolet (UV) light, and/or
- using combinations of these.

Where contact times are shorter than 10 minutes, usually in the range of 1–5 minutes, adequate mixing is a critical parameter. Mixing provides better dispersion of the disinfectant and forcing disinfectant contact with a greater number of microorganisms per unit time. The increased rate of collisions decreases the required contact time. Mixing can be accomplished by mechanical flash mixers at the point of disinfectant addition and at intermittent points or by specially designed plug flow contact chambers containing closely spaced, corrugated parallel baffles that create a meandering path for the wastewater (EPA 1977).

Chlorination/Dechlorination

This process is the extension of many present systems using Cl_2 gas or NaOCl. It incorporates controls and instrumentation to lower the Cl_2 dosage through measurement and limits Cl_2 residual to levels of 0.08 mg/l. The chlorinated effluents require sodium bisulfite feed (or another dechlorination process) and instrumentation to measure concentrations of sulfide, Cl_2 residual, and other indicators (e.g., pH and dissolved oxygen). Figure 4–22 is a typical schematic of the chlorination/dechlorination process including the equipment, tanks, and analytical measurement and control instrumentation. An issue to be addressed in the evaluation of this technique is pathogen occlusion by particles (see "Particle occlusion" later in this chapter) and regrowth in dechlorinated effluent.

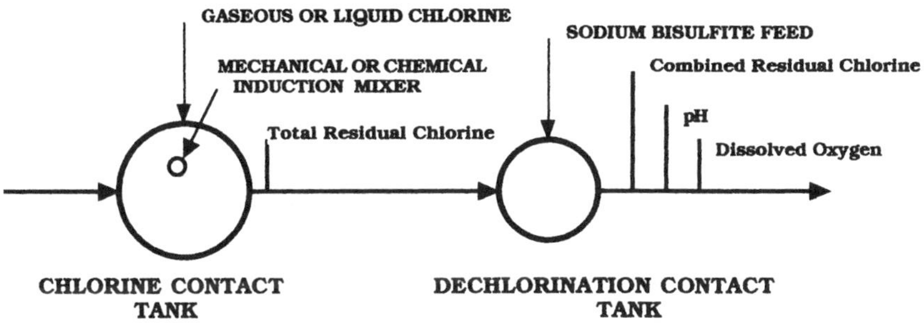

Figure 4–22. Chlorination/dechlorination.

Chlorine Dioxide

ClO_2 has historically been recognized as an outstanding disinfectant for the destruction of pathogens and the inactivation of viruses. ClO_2 has been demonstrated in EPA disinfection studies (Moffa et al. 1975; Drehwing et al. 1979) that have proven its disinfection ability. The key to its success has always been placed on bactericidal and viricidal effectiveness, onsite capabilities of generation, and control of byproducts. ClO_2, a yellowish gas at room temperature, was traditionally produced and used in an aqueous solution and is 10 times more soluble in water than Cl_2. Unlike Cl_2, ClO_2 does not react with NH_3 or other nitrogenous compounds, is not impaired over a wide pH range, and is effective longer. In water solutions it is soluble and stable up to 8,000 mg/l.

To obtain a supply on demand, a user employs a generation process designed to convert precursors to ClO_2. The most widely used precursor (outside of the paper and pulp industry, which generates 600 ton/d (540,000 kg/d) of ClO_2 from sodium chlorate) is sodium chlorite ($NaClO_2$). There are two basic types of $NaClO_2$-based ClO_2 generators: solution and gaseous.

The traditional solution-generating systems are represented by the acid, dissolved Cl_2 (NaOCl), and gaseous Cl_2 types, all of which rely on the use of sodium-chlorite solutions. The efficiency of generation as measured by the control or minimization of byproducts (free Cl_2) ranges from the acid system with the lowest efficiency to the gaseous Cl_2 system with the highest (95–98%). It normally takes a surplus of Cl_2 to complete the reaction in the generation of ClO_2 with these systems and consequently produces free Cl_2 to react and form other Cl_2 products, namely chlorates, chlorites, and hypochlorous acid. More recent technology generates ClO_2 in the gas phase, avoiding many of the byproduct issues associated with previous generation methods.

Gaseous generators use the following reaction of the chlorine gas/sodium chlorite system:

$$Cl_2\ (Gas) + 2NaClO_2\ (Solution) \longrightarrow 2ClO_2\ (Solution) + 2NaCl$$

The chlorine gas-sodium chlorite solution process for the generation of ClO_2 and treatment of wastewater is shown in Figure 4–23, and the UV light generation process is shown in Figure 4–24.

Sequential addition of Cl_2 followed by ClO_2 at intervals of 15–30 seconds enhanced high-rate disinfection beyond the expected additive effect (Moffa et al. 1975; Field et al. 1976). A minimum effective combination of 8 mg/l of Cl_2 followed 2 mg/l of ClO_2 was found as effective in reducing total and fecal coliform, fecal streptococci, and viruses to acceptable target levels as adding 25 mg/l Cl_2 or 12 mg/l ClO_2 individually. It was surmised that the presence of free Cl_2 in solution with chlorite ions (ClO^{-2}, the reduced state of ClO_2) may cause the oxidation of ClO^{-2} back to its original state. This process would prolong the existence of ClO_2, the more potent disinfectant.

Ozonation

O_3 is a highly reactive gas formed by electrical discharges in the presence of oxygen and is the most powerful oxidizing agent among present disinfection technologies. Because O_3 disinfects more rapidly than Cl_2, it requires shorter contact times and results in a lower capital cost for a contactor. It has the added benefit of supplying oxygen to the wastewater. O_3, does not produce chlorinated hydrocarbons or a long-lasting residual like Cl_2. O_3 is unstable and must be generated onsite just prior to application. Thus, unlike Cl_2, no storage is required.

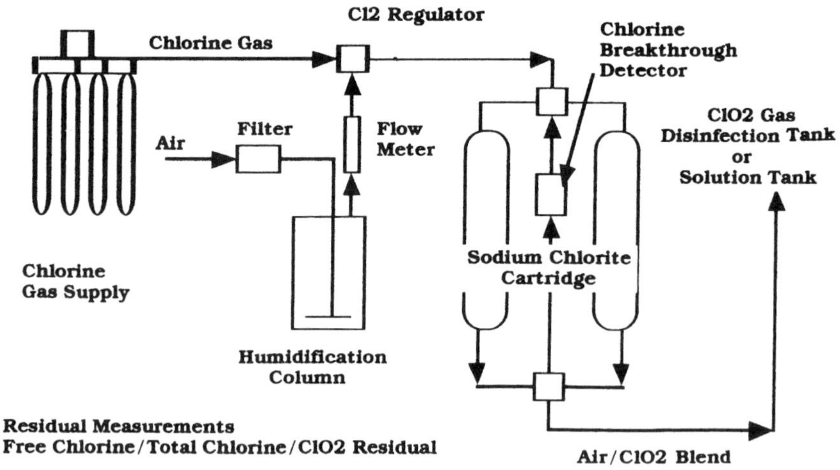

Figure 4–23. Chlorine gas-sodium chlorite solution generator of chlorine dioxide.

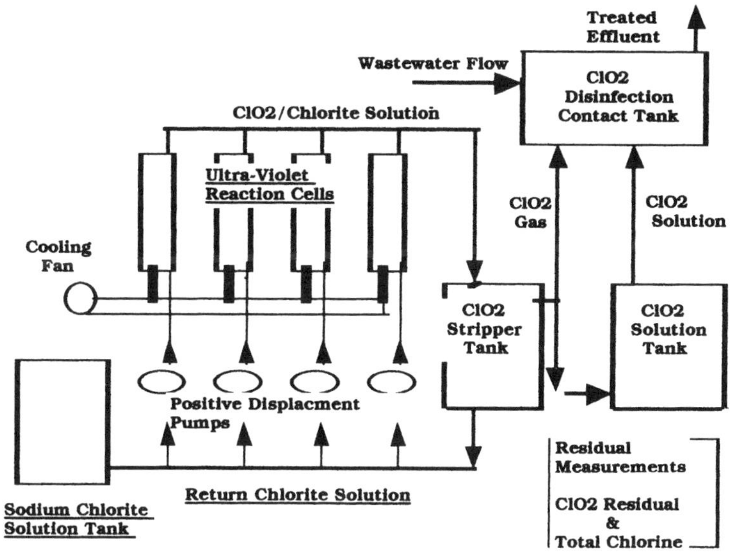

Figure 4–24. Ultraviolet-light generator of chlorine dioxide.

O_3's high level of chemical energy is the driving force for its decomposition and is a complex function of temperature, pH, and concentrations of organic and inorganic constituents of the wastewater stream. Because O_3 is an unstable gas and breaks into elemental oxygen rapidly, it requires onsite generation with an ozonator that applies an electrical charge across air or oxygen. A typical schematic is shown in Figure 4–25.

Wastewater ozonation may be accomplished by mechanical mixing, flow columns, diffusers, or an injector. This gas is applied to the wastewater being treated in a mixed contact chamber designed for efficient mass transfer to dissolve the O_3. Dosage and contact time requirements will vary with the wastewater characteristics but are usually 1–10 mg/l and <15 minutes, respectively.

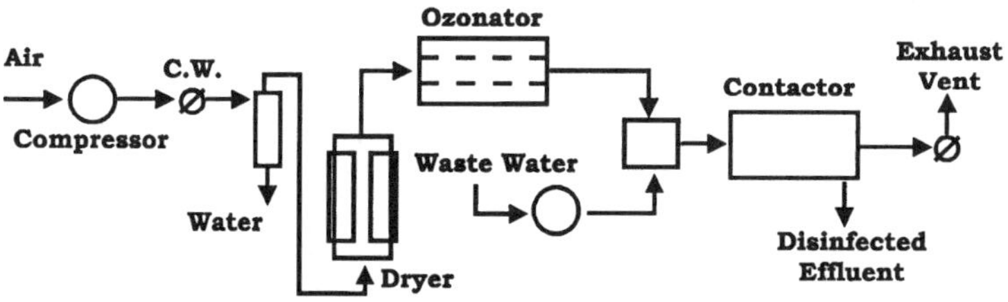

Figure 4–25. Typical ozone-treatment system.

O_3 systems have high capital and energy costs and produce byproducts (bromides).

Ultraviolet Irradiation

UV light at the germicidal wavelength of 254 nanometers (nm) disinfects water and wastewater by altering the genetic material (DNA) in cells so that bacteria, viruses, and other microorganisms can no longer reproduce. In UV disinfection systems, the UV light is produced by germicidal lamps that are submerged in either a closed vessel or an open channel. Thin film flows past the UV lamps and the microorganisms are exposed to a dosage of UV energy that takes 6–10 seconds. This dose is measured as the product of UV light intensity times the exposure time within the UV lamp array.

When UV light travels from its source through wastewater, suspended particles in the medium absorb the light. Specific applications take into consideration the expected level of microorganisms, SS concentration and size, the UV absorbing compounds in the wastewater, and the average and peak flowrates. This technique requires determining the effect of wastewater characteristics on UV performance, lamp fouling, and cleaning frequency. UV systems are designed and available in either horizontal or vertical arrays of lamps, depending on application and flow volume.

Facility Economics

Because of the characteristic intermittent operation associated with treatment of CSO, reduction of construction cost with a potential increase in operating costs often results in overall minimum costs. In the case of CSO chlorination facilities, the construction costs associated with contact basins having conventional contact time of 15–30 minutes are high and difficult to justify. Therefore, consideration should be given to higher mixing intensities and two-stage dosing to make better usage of the Cl_2 and/or higher Cl_2 dosages and smaller, shorter-detention-time contact basins to effect the same microorganism kill. Also, the use of more rapid oxidants (disinfectants) than Cl_2 (e.g., ClO_2, O_3, and UV irradiation) can result in a more economical disinfection facility.

Disinfectant costs for separate CSO treatment are higher than those for sanitary wastewater. This is the result of smaller total annual disinfectant volume requirements, increased disinfectant concentration requirements, and higher unit O&M costs for CSO. These costs could be reduced by using the facilities in conjunction with DWF treatment plants, whenever possible.

Curves comparing generation and feed costs for Cl_2 gas, ClO_2, and hypochlorite generation disinfection systems for CSO are presented in Figure 4–26. These costs include equipment, labor, piping, housing, electrical and instrumentation, and miscellaneous items. No allowance for land was included.

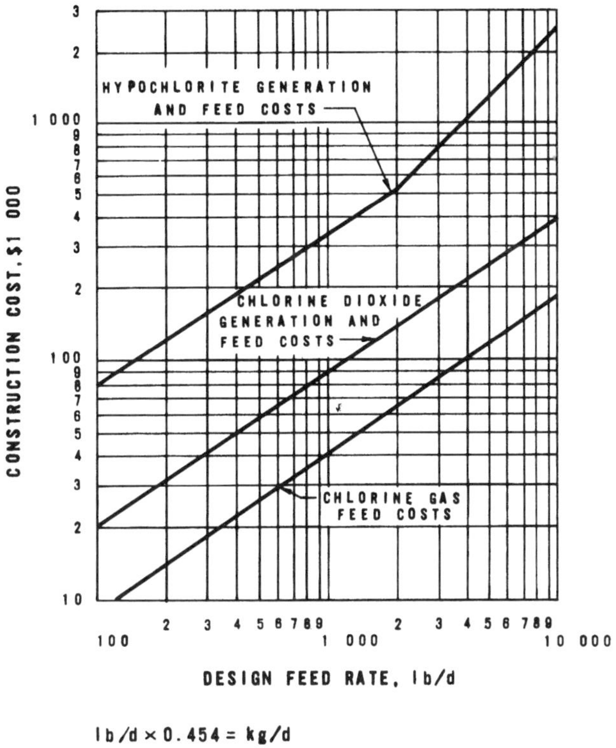

Figure 4–26. Chlorine-disinfection-feed facilities cost curves (ENR = 2000) (Chlorine gas costs do not include dechlorination.) (Lager et al. 1977).

Capital and operating costs for several CSO and stormwater disinfection facilities are presented in Table 4–16. As previously mentioned, conventionally long contact times may not be economical. Short contact times with more intense mixing, using a basin and mixer similar to those used in coagulant mixing, can effect the same disinfection results. Construction costs curves for high-intensity mixing Cl_2-contact basins are presented in Figure 4–27. Power requirement curves for high-intensity mixing are presented in Figure 4–28.

Particle Occlusion

Disinfection by chlorination or other oxidants is often effective only for the free-floating, near-surface, or surface microorganisms. This is especially true for high-rate disinfection of WWF due to the relatively short contact times. Depending on the degree of upstream treatment (SS removal), the disinfection process may not be effective if microorganisms are contained or occluded in the intersti-

Table 4–16. Cost Data on Chlorine Gas and Hypochlorite Disinfection (ENR = 5,000) (Lager and Smith 1974)

Location, Agent, Source	Capital Cost ($)	Operating Cost ($/yr)	Cost/lb Available Chlorine ($)
Akron, OH*			
Sodium hypochlorite			
Purchased	1,104,000	58,200	0.37–0.67
Cambridge and Somerville, MA*			
Sodium hypochlorite			
Purchased	—	—	0.97
On-site generation	—	—	0.51
New Orleans, LA†			
Sodium hypochlorite			
On-site generation	1,448,000	724,000	0.30
Saginaw, MI‡			
Chlorine gas	406,000	5,800	0.87
Sodium hypochlorite			
Purchased	49,000	16,000–30,000	0.45–.78
On-site generation	239,000–403,000	11,700–13,000	0.69–1.0

(continued)

South Essex Sewage District, MA§			
Chlorine gas	1,667,000	582,000	0.09
Sodium hypochlorite			
Purchased	1,054,000	913,000	0.12
On-site generation			
Seawater	4,200,000	406,000	0.09
Brine	4,200,000	758,000	0.13

*CSO disinfection.

†Storm sewer discharge disinfection.

‡CSO disinfection at use rate of 42,000 lb/yr of chlorine.

§Sewage treatment plant effluent disinfection at use rate of 24,000 lb/day of chlorine.

$/lb X 2.2 = $/kg

(EPA-670/2-74-040)

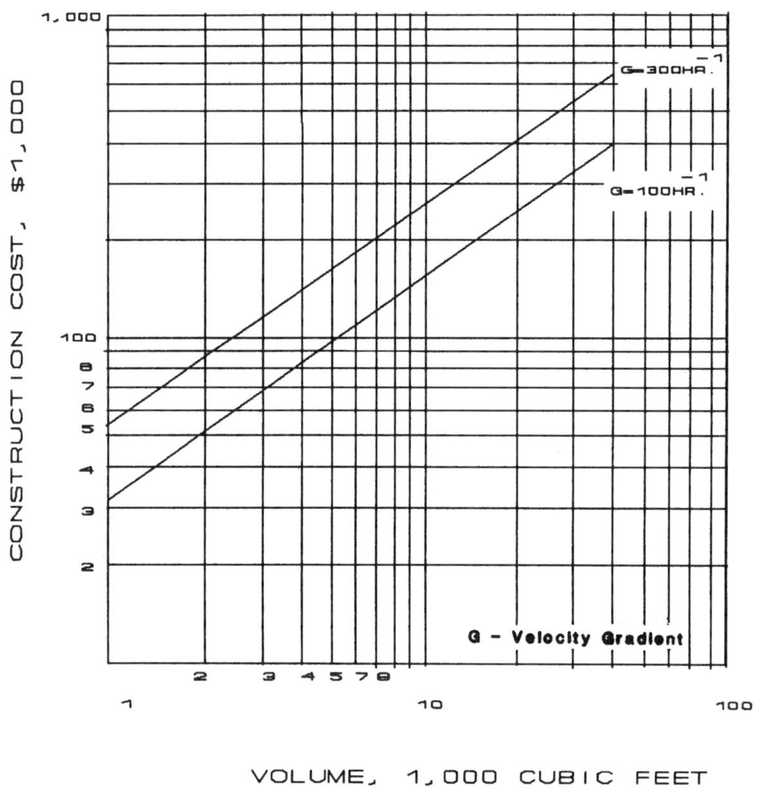

Figure 4-27. High-intensity mixing/chlorine-contact-basin cost (Benjes and Culp 1976).

cies of larger protective solid particles and fluid droplets. Chlorination and disinfection processes have limits relative to chemical penetration, which is a function of chemical demand ("chlorine demand") and microorganism occlusion by particles. These limits are based on the concentration, size, make-up, and morphology of wastewater particles; water chemistry; contact time; mixing; and disinfectant dosage/intensity. Therefore, determination of particle characteristics is necessary for WWF disinfection process design.

The adverse impacts associated with chlorination are among the issues leading to the development of alternative methods of disinfection. These methods can have problems similar to those of Cl_2 disinfection. Specifically, disinfection by UV irradiation of effluents containing high concentrations of relatively large protective particles and droplets has proven to be ineffective. The absorption of UV radiation by these substances attenuates the available UV energy and reduces the depth of penetration into the wastewater (Roeber and Hoot 1975). This reduction in disinfection effectiveness is further significantly compounded by mi-

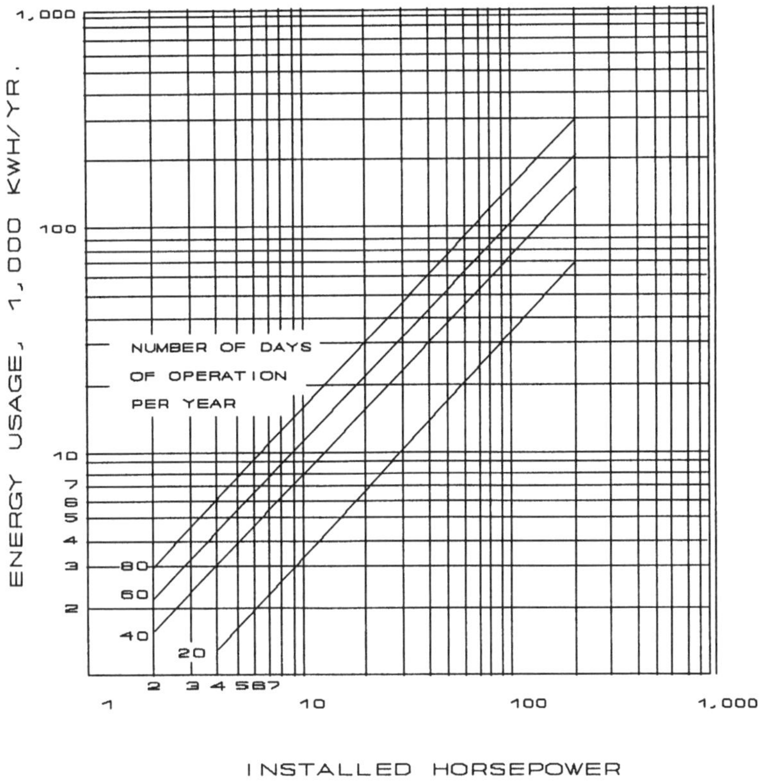

Figure 4-28. High-intensity mixing (chlorine contact) power requirements (hp 3 0.746 5 KW).

croorganisms within relatively large particles experiencing little or no radiation as a result of the absorption of the radiation by the outer, protective layer.

Optimization

CSO must be controlled by a storage-treatment system because storm flow in the combined sewerage system is intermittent and highly variable in both pollutant concentration and flowrate. A treatment facility operating without the benefit of upstream storage would need to be very large and costly in order to handle the relatively high volume and flowrate of a CSO. Similarly, if storage is used without treatment, the storage volume required would be very large and also expensive.

This optimized system maximizes the use of the existing system before new construction and sizes the storage volume in concert with the WWTP capacity to obtain the lowest cost storage and treatment system.

One commonly used approach to CSO pollution abatement is to rely on a storm-event based design of storage-tank volume to capture CSO for pump-back and/or bleed-back (gravity-flow) to the existing WWTP for treatment. However, this may not be the most economical approach because of the high costs involved in the construction of storage and/or expanded WWTP capacity. Therefore, a strategy should be adopted to create an optimized CSO control system that maximizes the use of the existing system before constructing new facilities and sizes the storage volume together with the treatment rate to obtain the lowest-cost storage-treatment system. An integrated optimization strategy will minimize the construction of new storage and treatment facilities by:

- making operational and low-cost, inline improvements in the existing combined sewerage (sewer-treatment) system to maximize storage and treatment before new storage or treatment unit construction;
- accounting for the dual function of treatment by sedimentation when storage tanks overflow;
- maximizing the existing WWTP capacity;
- selecting design capacity of the CSO control system to reflect the most economical approach by choosing the point of diminishing returns on the curve of pollution control vs. cost; and,
- sizing the storage-treatment system according to the most economical break-even point between the amount of storage vs. the treatment capacity.

Improving Analysis

To effectively employ control technologies in a combined sewerage/drainage system, the control functions, applicability, and idiosyncrasies of their individual designs must be clearly understood. Two major problems that continue to occur and that hinder the performance and reputation of various control technologies are:

- inadequate or faulty flow sampling and measuring techniques for proper design and efficiency evaluation; and
- inappropriate application or placement of controls in the sewerage system.

The monitoring and analyses needed for proper application and assessment of CSO control technologies are expensive and complex in the short term; however, reliable data collection may save even more expensive construction costs by eliminating unnecessary facilities and/or additional controls. Sampling devices must be able to capture the heavier SS or settleable solids and not manifest biased results due to stratification. For an automatic sampling device, this means that its intake velocities and ports must be greater than the mainstream velocity and be placed at multiple levels, respectively, in order to capture the heavier particles near the channel invert.

Because the SS-settling-velocity distribution is relatively constant for sanitary wastewater, the fundamental hydraulic loading or overflow-rate design parameters previously established for sanitary wastewater primary settling tanks cannot readily be used for CSO and, therefore, site-specific solids characterization of CSO is necessary for the satisfactory design of physical treatment (e.g., sedimentation, vortex separation, and fine-mesh screening or microscreening). Sedimentation and vortex separation treatability is dependent upon the (1) fraction of settleable solids and SS with respect to the total solids, (2) SS-settling-velocity distribution, and (3) hydraulic loading (gal/ft^2/min). Fine-mesh screening and microscreening treatability is dependent upon the (1) SS-particle-size distribution, (2) fraction of filterable SS with respect to total solids, and (3) hydraulic loading.

It is important to characterize CSO SS on a site-specific basis because the build-up of settleable solids in the sewer system is a function of local conditions that include topography, geology, and antecedent dry-weather period. The topography influences the sewer slope or gradient with milder slopes, causing greater amounts of solids deposition that are subsequently resuspended during the more intensive storm flows. Also, the surrounding geology, or more specifically the soil characteristics, affects the suspended-/settleable-solids concentration and particle-settling-velocity distribution.

Furthermore, suspended-/settleable-solids concentrations will vary with time during the storm. First flush commonly refers to the segment of flow with a high resuspended sediment concentration at the beginning of a storm. However, first flush should not be used as a design principal because this concentrated sediment concentration of the flow is dependent on the sewer system, SS, and rainfall characteristics and therefore may come at any time during a storm, if it comes at all. These storm-flow variations require that sampling be done for the duration of the storm and for several storms in order to develop a long-term average of the solids concentration and particle-settling-velocity distribution.

After samples have been collected and analyzed for SS, two particle-settling characteristic analyses should be conducted: one for settleable solids and the other for settling-velocity distribution. These analyses will enable a site-specific estimate of the percent of SS that the intended CSO control facility is capable of removing.

The settleable-solids analysis should be the gravimetric type with data presented in mg/l (Standard Methods 1992: Section 2540F) to determine the fraction of settleable solids in the storm flow. The method uses a column of at least 8 inches (20 cm) in depth. A sample is siphoned from the center of the column after 1 hour of quiescent settling to determine the nonsettleable solids. Settleable solids are equal to the initial SS concentration minus the nonsettleable-solids concentration. However, this method may not be applicable to control technologies with detention times <1 hour (e.g., swirl/vortex devices).

A crude method of estimating the fraction of SS that swirl/vortex devices can concentrate (i.e., SS with a settling velocity greater than 0.10 cm/sec) is to siphon a sample at mid-depth after 100 seconds. The SS that settle during the first 8 seconds of the test (visual observation) are the major portion of the SS that the swirl was developed to remove (based on a settling velocity of 2.6 cm/sec or greater for an 8-inch [20 cm] column).

Two methods specifically developed for determining particle-settling-velocity distributions of swirl and vortex application are:

- Brombach or German (Michelbach and Wöhrle 1993) and
- Norwegian Institute for Water Research (NIVA) (Lygren and Damhaug 1986; Walker et al. 1993).

These methods are specifically designed to determine the concentration of heavier particles in storm-generated flows (the high settling-velocity particles that the swirl/vortex were developed to remove) and, accordingly, offer significant benefits over traditional settling-column methods that relate to sedimentation treatment. These newer methods require less sampling (one sample from the bottom as opposed to multiple-simultaneous samples from side ports) and smaller testing volumes (approximately 1–3 gallons [4–12 l] as compared to approximately 10–15 gallons [38–57 l]). They use a smaller apparatus that is more amenable to field use (less than 3.3 feet [1 m] deep and 2 inches [5 cm] wide as compared to approximately 6–8 feet [1.9–2.7 m] or deeper and 8–10 inches [20–30 cm wide]). These newer methods provide truer representation of high settling-velocity SS because a concentrated sample is dropped into the settling column at starting time, t_0. It is difficult to have a homogeneously mixed sample at t_0 using a traditional settling column that results in lower-than-actual predictions of settleable-solids fractions.

There are other new methods available for high-rate particle-settling velocity analysis. Because there is currently no standard, it is up to the engineer to determine the most appropriate method to match high-rate treatment design objectives. However, caution is warranted as current literature (Aiguier et al. 1995) suggests that the derived settling-velocity curves from the various methods tend to give different results.

Economic Analysis

A complete and thorough storage-treatment system analysis for the entire drainage area should be conducted using break-even economics for cost optimization of the overall pollution-control requirements of the system. Storage-treatment strategies should identify the point of diminishing returns based on long-term rainfall-CSO quantity and quality analyses, identified by the increasing cost without increased benefits to treatment. Figure 4–29 contains site specific data for Atlanta, Georgia (Heaney et al. 1977), and shows that the rate of increase in the overall percent precipitation control drops off dramatically after the 90% control point. This point should be selected for design of the system unless

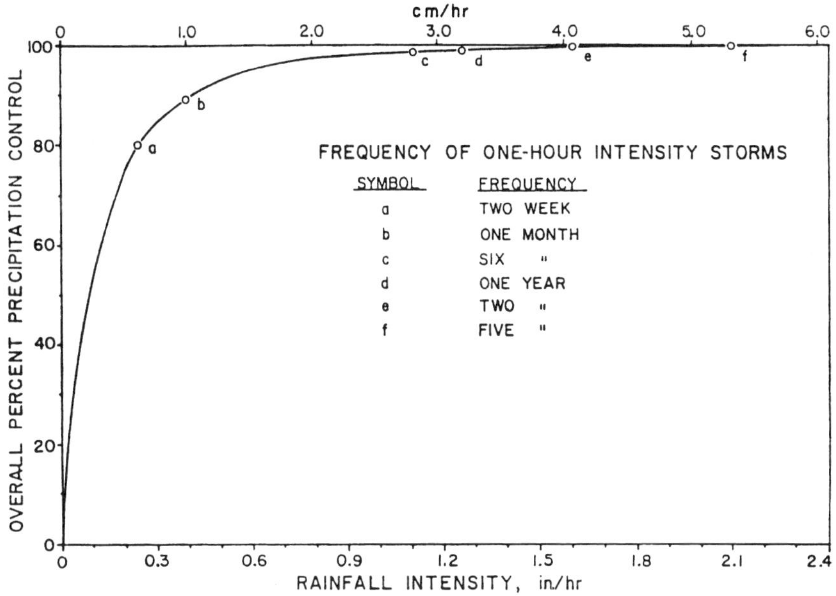

Figure 4–29. Overall percent precipitation control versus rainfall intensity, Atlanta (Heany et al. 1977).

regulations or receiving water impacts require a higher capacity system. The exact shape of this curve will vary from site to site.

Another aspect of system optimization is illustrated by the classic economic optimization in Figure 4–30. This example demonstrates that for a certain design load the solid bar in the upper chart can handle all combinations of storage and treatment and that the construction cost for a single storage tank or a treatment facility able to handle this flow separately increases. However, an optimized storage-treatment system design for this event (the dashed line in the lower figure) can handle this CSO event at a significant cost savings of either the storage or treatment options taken separately. The storage and treatment design combinations for this example demonstrate that in general cost optimization is achieved only when the storage-treatment system does not rely too heavily on either storage or treatment.

Maximizing WWTP Treatment Capacity

Approaches to maximizing WWTP capacity include:

- operational changes
- lower cost modifications
- retrofitting of primary settling tanks and
- parallel processes.

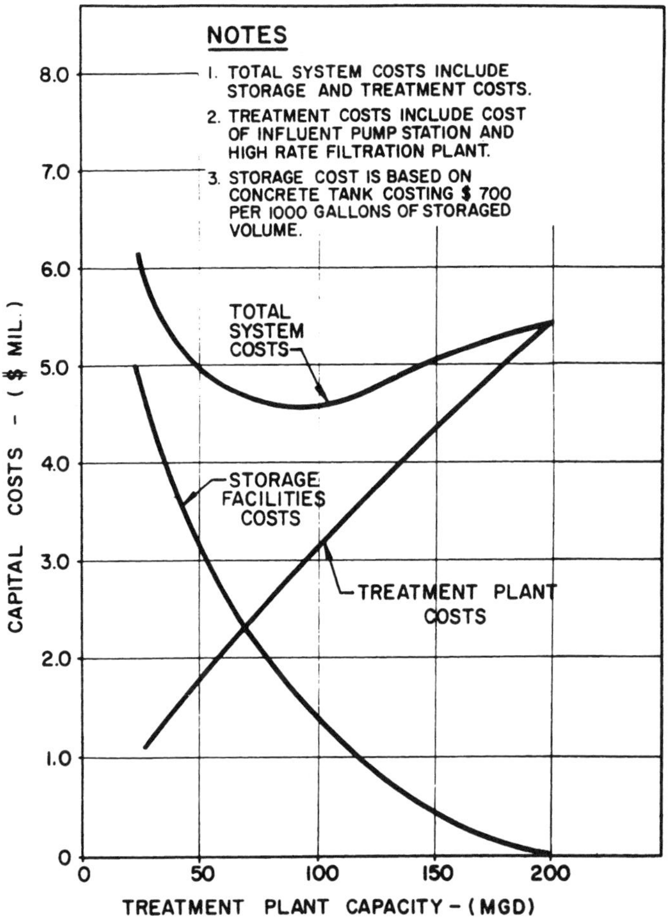

Figure 4-30. Estimated capital costs of storage and treatment for 200 MGD overflow (from *The Control and Treatment of Industrial and Municipal Stormwater,* P. Moffa, ed.).

Operational Changes

Advantage should be taken of the storage available in the existing collection system (EPA 1995a). Storage capacity exists in most sewer lines because they were designed for the very infrequent (e.g., 1 in 5 and 10 years) storms for flood control. Use of tanks at the WWTP or abandoned treatment works (i.e., abandoned because they are incapable of achieving secondary effluent standards yet are fully capable of achieving primary treatment) should also be considered.

At capacity, overflowing storage tanks behave like primary settling tanks. This dual-function of sedimentation treatment obtained during overflows should be taken into account before any further control designs. Removal by settling can be estimated by continuous long-term modeling of storm-flow influent to the storage tank using the settling velocity distribution of the influent particles and the overflow rate of the tank (Driscoll et al. 1986).

In optimizing the WWTP, the first and cheapest optimization alternative is to direct more flow to the WWTP and through the primary settling tanks. This entails maximizing the treatment at the WWTP during storm-flow conditions by effectively treating as much flow as possible during the storm-flow occurrence (Field and Struzeski 1972; EPA 1995a and b).

This practice is well worth the effort. It reduces the release of raw sanitary sewage and polluted storm water at upstream outfalls by redirecting the CSO to the WWTP, where at least some measure of primary treatment can be accomplished. In some systems for certain portions of the storm-flow period, the particle settling velocities are relatively high in comparison to the DWF particle-settling velocities due to the resuspension of antecedent DWF deposits in the sewer system. Therefore, during storm-flow periods, higher flowrates can be directed through the primary settling tanks.

However, decreases in settling tank efficiency as a function of the higher influent hydraulic loading or overflow rate should be taken into account. As the WWTP flowrate (or primary-settling-tank-overflow rate) increases, its efficiency will decrease until a point is reached after which diverting more flow to the WWTP will have a negligible effect on the increase of CSO pollution removal. If the increased hydraulic loading will cause a serious upset of a secondary biological process, then primary treatment plus disinfection, barring regulatory considerations, should be used as a minimum measure (EPA 1995a).

Catchbasin cleaning was proven to offer cost effective maintenance and solids handling reduction at a WWTP in a study conducted in Boston (Aronson et al. 1983). However, in this case study, it was found that an operational commitment to cleaning the catchbasins a minimum of twice a year was required to achieve cost-effective pollution control.

Lower Cost Modifications

Treatment maximization at the WWTP can also be enhanced by using advanced signals that indicate relatively high storm flows from monitoring systems (comprised of flowmeters, raingauges, and/or radar) located upstream of the WWTP. Two methods are:

- maximizing the use of insewer storage by simple CSO-flow regulator adjustments (Feuerstein and Maddus 1976). Table 4–17 shows the cost benefit for this study, as storage was doubled without additional cost; and,
- installing dynamic regulators that respond to remote-sensing devices to increase insewer storage (Leiser 1974; Watt et al. 1975).

Table 4-17. Spring Creek Auxiliary Water Pollution Control Plant, New York, NY (from Lager et al. 1977b)

	Storage capacity in Mgal	Drainage area in acres	Capital cost in $	Storage cost in $/gal	Cost per acre in $/acre	Annual operation & maint. cost in $/yr
Storage	12.39	3,260	29,837,780	2.40	9,125	250,580
Sewer	13.00	—	—	—	—	—
Total	25.39	3,260	29,837,780	1.17	9,125	250,580

In conjunction with directing more flow to the WWTP, an evaluation should be made to increase the interceptor-carrying capacity from storage (insewer and/or tank) to the WWTP. This evaluation should include the use of:

- increased pumping capacity for surcharged interceptors (Watt et al. 1975), which is the easiest and cheapest method, as it is an operational change that does not require additional capital cost; and
- polymer injection, which has been effectively demonstrated to reduce sewer roughness and increase carrying capacity in both open-channel (Derick and Logie 1973) and surcharged flows (Chandler and Lewis 1977).

Retrofitting of Primary Settling Tanks

The performance of the primary settling tanks can be enhanced by retrofitting with appurtenances that improve settling and/or separation of SS so that higher overflow rates and shorter detention times can be used. Proven methods are:

- converting to higher rate DAF. Although this system has higher operational and maintenance costs than settling, it can be automated to operate only during storm-flow conditions, avoiding additional power costs during periods of DWF. In Clatskaine, Oregon (Benedict and Roelfs 1981), a DAF-contact stabilization plant successfully treated storm-induced sanitary-sewer overflows at a capital cost of only 14% greater than a standard DWF WWTP with only 10% increased total cost (including O&M);
- installing plate or tube settlers. Lamellae-plate settlers, which are installed on an angle, increase the effective settling area of the tank. Demir (1995) found that the best sedimentation efficiency occurred at plate angles of 50° over a spread of hydraulic loadings 0.2–1.2 gal/ft^2/min (0.5–3.0 m^3/m^2/h). Through modeling and experimental design, Fadel and Baumann (1990) determined that steeply inclined tube settlers (45°–60°)

operated best when restricted to a 40% reduction in depth due to sludge accumulation before backwashing.

- employing chemically enhanced primary treatment (CEPT), which uses metal salts and polyelectrolytes. Morrissey and Harleman (1992) cite several successful application of polyelectrolytes reducing the use of metal salts. Primary and secondary sludge for the Hyperion Treatment Plant of San Diego, California, increased only 10% and 6%, respectively, before and after digestion. CEPT at Hyperion attained removals of 83% SS and 52% BOD_5 at an average hydraulic loading of 1.2 gal/ft^2/min (3 m^3/m^2/h) with additions of 20 mg/l of ferric chloride and 0.25 mg/l of anionic polymer; and
- installing carrier-enhanced-settling systems as previously described under the "Treatment" section.

The EPA swirl degritters or the HIL Grit King (with bottom hoppers for storing detritus, as opposed to swirl regulators, which have a continuous foul-effluent flow) could replace conventional aerated degritters. Vortex degritters have no moving parts and require less space than conventional grit chambers. Vortex degritters and/or swirl/vortex separators (with underflow to the treatment plant) can be constructed at satellite facilities near CSO points or at the WWTP upstream and in series with other storage/treatment methods. Present worth analyses of the swirl degritter versus aerated grit chambers show savings of 15% for a unit capacity of 1.0 Mgal/d (3,800 m^3/d) and 33% savings for a 10 Mgal/d (38,000 m^3/d) unit (Sullivan et al. 1982).

Another innovative technology suitable for retrofitting at the treatment plant, in the conveyance system or as satellite treatment, is the previously mentioned CDS.

Parallel Processes

Parallel facilities at the WWTP might consist of high-rate fine-mesh screens or microscreens, high-rate filtration, or relatively slow-rate settling tanks. Parallel interceptors to the WWTP could also be constructed to increase in-sewer storage.

High-rate physical/chemical treatment processes have shown their importance in the treatment of CSOs. These physical processes, with or without chemical enhancement, can be designed for automated operation, are adaptable to intermittent and variable quantity and quality characteristics, and are resistant to shock loadings. They can either be incorporated into the existing WWTP as parallel processes for storm flow or set up as part of remote satellite control plants in conjunction with storage facilities at outfall sites.

These high-rate methods, along with the previously mentioned retrofitting processes can also be used in series during the DWF periods to improve overall plant efficiency.

Storage

An important part of CSO control-system planning is the determination of the economic break-even point between the required storage volume versus treatment capacity. Figure 4–31 shows that there is an increase in BOD_5 and SS percent removals, with an increase in tank volume per drainage area. However Figure 4–32 demonstrates decreasing removal efficiencies per unit volume as tank size increases. Beyond an optimum tank volume (knee-of-the-curve), the rate of cost increases more rapidly than the benefit for retaining the extra flow increases; therefore, it is not economical to design storage facilities for the infrequent storm.

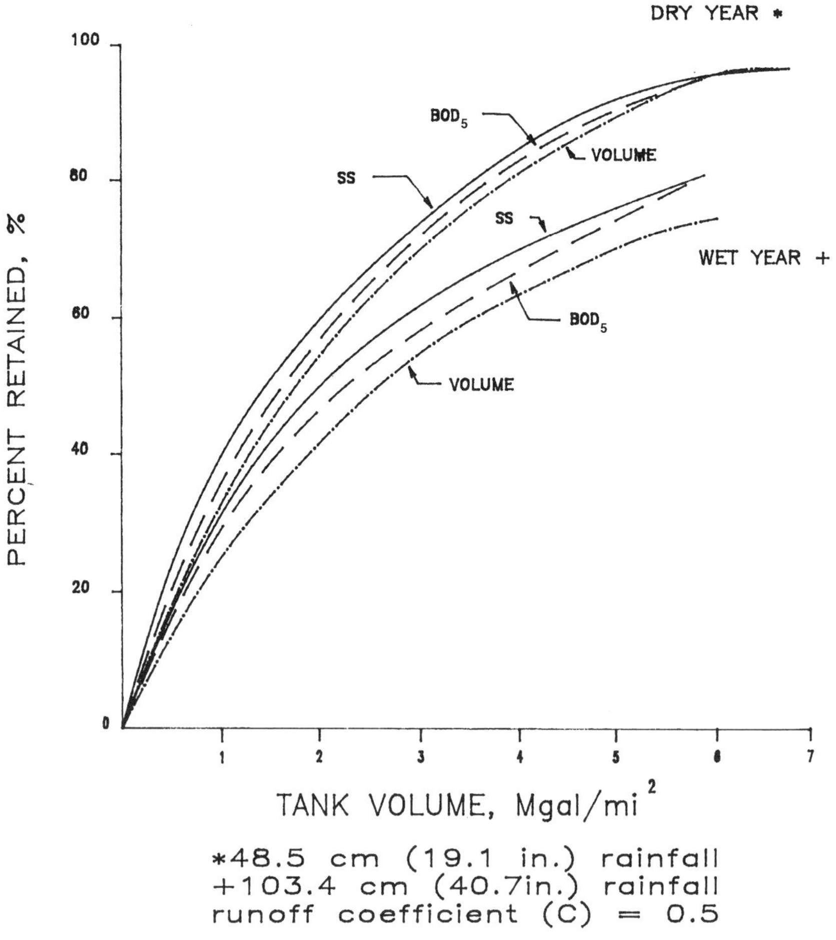

Figure 4–31. Percent retained versus tank volume (Lager et al. 1977b).

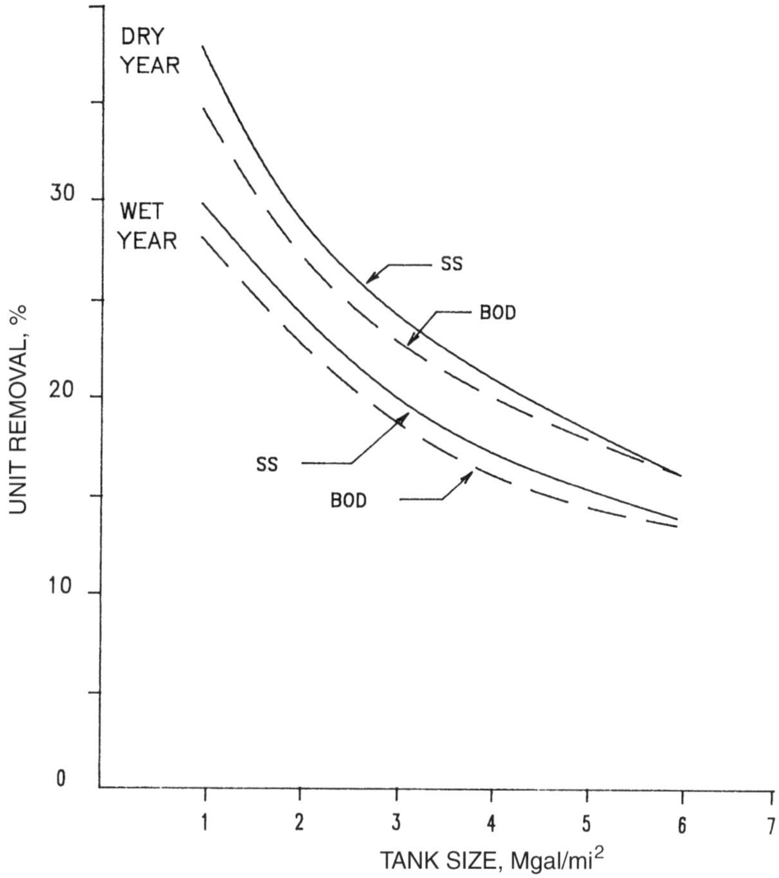

Figure 4-32. Unit removal efficiencies for CSO detention tanks (Lager et al. 1977b).

Satellite Treatment

Satellite treatment should consist of automatic physical processes with remote monitoring to reduce manpower requirements. Maintenance issues will be more pronounced, and potential failures at remote locations along with associated water-quality impacts will exist. Location issues such as sensitive receiving-stream segments and neighborhood aesthetics should be considered. Accordingly, physical treatment processes at satellite locations should be used only if the WWTP cannot handle all of the WWF. As previously stated, satellite storage should be used for dual-purpose treatment by settling for WWF volumes greater than the storage volume.

New Sewerage Systems

New urban areas or upstream additions to older combined sewer systems (CSS) should use advanced designs requiring larger diameter sewers to add storage capacity to the system; in addition, steeper sloped sewers with more effective bottom cross-sections will eliminate antecedent DWF pollutant deposition and resulting concentrated storm flushes (Kaufman and Lai 1978; Sonnen 1977). The additional capital cost of an advanced CSS would be incrementally small considering the:

- overall cost of installing a conventional CSS or a two-pipe (storm and sanitary) separate sewer system, and
- cost effectiveness for storm-flow pollution control.

The design of new WWTPs should include treatment of CSO and not just treatment for peak DWF conditions. Larger interceptors, higher treatment flowrates, and alternative high-rate treatment methods, as previously mentioned, should also be used in new WWTP designs.

Analysis

Any approach to CSO pollution abatement should include operational changes, long-term modeling, and low-cost modifications before higher cost storage and treatment construction. Figure 4–33 provides a flowchart summarizing this approach. However, each collection-treatment system will require its own optimization plan, as no one method of storage, treatment, or O&M plan has been demonstrated as a cure-all. Only through evaluation of the total system can the appropriate methodology and technology be chosen for CSO pollution-abatement system optimization.

CSO pollution abatement must be dealt with in terms of a storage-treatment system because storm flows are intermittent and highly variable. Construction of additional treatment facilities or storage tanks to control CSOs is expensive, and therefore the optimization strategy must:

- maximize the available capacity of the existing sewerage (collection-treatment) system;
- include the benefit of treatment by settling in overflowing storage tanks;
- size the storage-treatment system based on continuous long-term historical, urban hydrology, and knee-of-the-curve for economical analysis (Figure 4–29);
- size the storage-treatment combination using a break-even point or low-cost economical analysis (Figure 4–30); and,

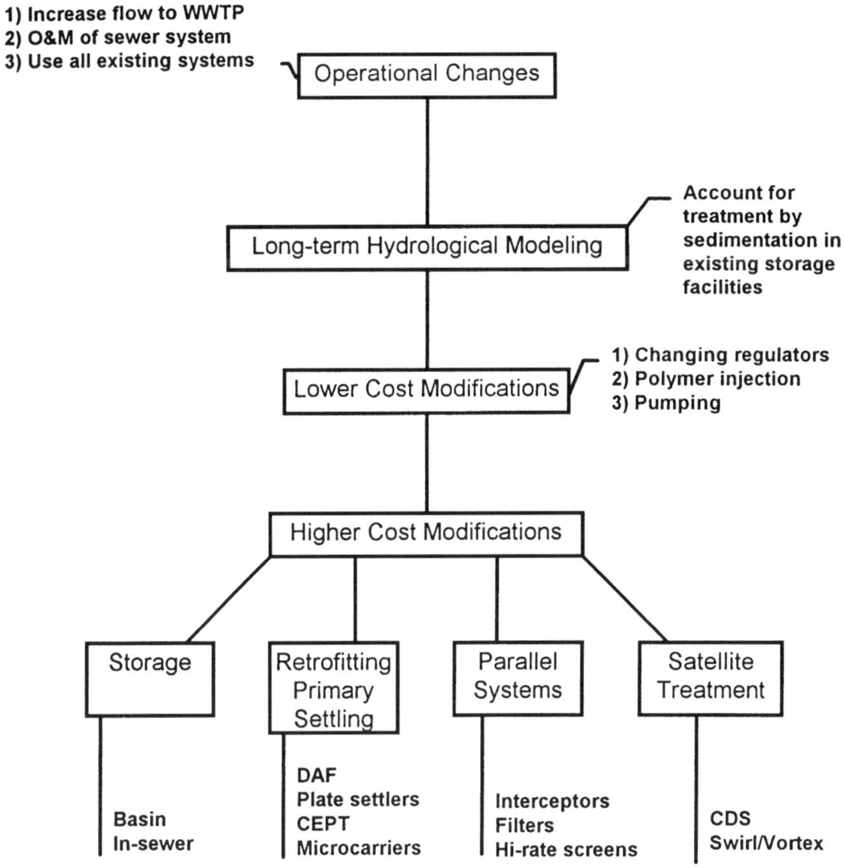

Figure 4–33. Flowchart for optimization.

- use a mass-flow analysis in the selection of control capacity, which does not assume storage tanks will be empty prior to the next storm-flow occurrence (Figures 4–31 and 4–32).

The existing WWTP will have treatment limitations due to the interceptor capacity (which is analogous to a bottleneck), expansion costs, and space availability. Therefore, besides the enhancement of capacity for the interceptor and facilities at the WWTP, the storage-treatment system optimization alternatives should include high-rate treatment with storage or satellite facilities at an upstream CSO point.

Acknowledgments

The authors would like to thank these two people for their significant contributions to the updating and technology description of this second edition:

William C. Pisano, Ph.D., P.E., of Montgomery Watson, Inc., for specific information and writings on innovative regulators, sewer and tank flushers, and the Fluidsep vortex device.

Frank Rogalla of Metcaff and Eddy, Inc., for specific information and writing on biological aerated filters (BAF).

Research was supported in part by an appointment to the postgraduate research program at the National Risk Management Research Laboratory, Cincinnati, Ohio, EPA-administered (NRMRL/EPA) by the Oak Ridge Institute for Science and Education through an interagency agreement between the U.S. Department of Energy and NRMRL/EPA.

References: Chapter 4

Agnew, R. W., et al. 1975. "Biological treatment of combined sewer overflow at Kenosha, Wisconsin." EPA, Cincinnati, OH, Rep. EPA-670/2-75-019 (NTIS PB 242 126).

Aiguier, E., G. Chebbo, J. L. Bertrand-Krajewski, P. Hedges, and J. N. Tyack. 1995. "Methods for determining the settling velocity profiles of solids in storm sewage." International Conference on Sewer Solids: Characteristics, Movement, Effects and Control. Dundee, Scotland, IAWQ/IAHR.

Alleman, J. E., J. E. Etzel, D. E. Gendron, J. C. Conley, W. F. MaCoy, and A. J. Hein. 1986. "Bromine-based disinfection performance." Presented at the 61st Annual Conference of the Water Pollution Control Federation, Dallas, Texas, October 1986.

American Public Works Association. 1970a. "Combined sewer regulator overflow facilities." Federal Water Quality Administration, Rep. 11022DMU07/70 (NTIS PB 215 902).

———. 1970b. "Combined sewer regulation and management—A manual of practice." Federal Water Quality Administration, Rep. 11022DMU08/70 (NTIS PB 195 676).

Aronson, G. L., D. S. Watson, and W. C. Pisano. 1983. "Evaluation of catchbasin performance for urban stormwater pollution control." EPA, Cincinnati, OH, Rep. EPA-600/2-83-043 (NTIS PB83-217745).

Benedict, A. H., and V. L. Roelfs. 1981. "Joint dry-wet weather treatment of municipal wastewater at Clatskanie, Oregon." EPA, Cincinnati, OH, Rep. EPA-600/2-81-061 (NTIS PB81-187262).

Benjes, H. H., and W. Culp. 1976. "Cost estimating manual—Combined sewer overflow storage and treatment." EPA, Cincinnati, OH, Rep. EPA-600/2-76-286 (NTIS 266-359).

Brombach, H. 1986. "Funktion und Bemessung von Luftregulierter Heber," Umwelt- und Fluid-Technik GmbH, Bad Mergentheim, Germany.

———. 1987a. "Funktion und Bemessung Spuelkippen," Umwelt- und Fluid-Technik GmbH, Bad Mergentheim, Germany.

———. 1987b. "Liquid-Solid Separation at Vortex-Storm Overflows." Fourth International Conference on Urban Storm Drainage, Lausanne, Switzerland.

———. 1992. "FluidBend Produktinformation," Umwelt- und Fluid-Technik GmbH, Bad Mergentheim, Germany.

Chandler, R. W., and W. R. Lewis. 1977. "Control of sewer overflows by polymer injection." EPA, Cincinnati, OH, Rep. EPA-600/2-77-189 (NTIS PB 272 654).

Chebbo, G., et al. 1995. "Technical solutions envisaged in managing solids in combined sewer networks." International Conference on Sewer Solids—Characteristics, Movement, Effects, and Control. Dundee, Scotland. IAWQ/IAHR.

Demir, A. 1995. "Determination of settling efficiency and optimum plate angle for plated settling tanks." *Water Research,* IAWQ 29(2).

Derick, C., and K. Logie. 1973. "Flow augmenting effects of additives on open channel flows." EPA, Cincinnati, OH, Rep. EPA-R2-73-238 (NTIS PB 222 911).

Drehwing, F., A. J. Oliver, D. A. McAurthur, and P. A. Moffa. 1979. "Disinfection/treatment of combined sewer overflows." EPA, Cincinnati, OH, Rep. EPA-600/2-79-134 (NTIS PB80-113459).

Driscoll, E. D., D. DiToro, D. Gaboury, and P. Shelley. 1986. "Methodology for analysis of detention basins for control of urban runoff quality." EPA, Washington, D.C. Rep. EPA-440/5-87-001 (NTIS PB87-116562).

EPA, 1973. "Combined sewer overflow seminar papers." EPA, Cincinnati, OH, Rep. EPA 670/2-73-077 (NTIS PB 231 836).

———. 1995a. "Combined sewer overflows: Guidance for nine minimum controls." EPA, Washington, D.C. Rep. EPA-832-R-95-003.

———. 1995b. "Combined sewer overflows: Guidance for long-term control plan." EPA, Washington, D.C. Rep. 832-B-95-002.

Fadel, A. A., and E. R. Baumann. 1990. "Tube Settler Modeling." *Journal of Environ. Eng.* ASCE, 116(1): 107–24.

Feuerstein, D. L., and H. F. Maddus. 1976. "Wastewater management program," vol. 1, summary report. EPA, Cincinnati, OH, Rep. EPA-600/2-76-222a (NTIS PB 260 887).

Field, R., and E. J. Struzeski. 1972. "Management and control of combined sewer overflows." *Journal of Water Pollution Control Federation,* vol. 44, no. 7 (July 1972): 1393–1415.

Field, R., V. P. Olivieri, E. M. Davis, J. E. Smith, and E. C. Tiffit. 1976. "Proceedings of workshop on microorganisms in urban stormwater." EPA, Cincinnati, OH, Rep. EPA-600/2-76-244 (NTIS PB 263 030).

Field, R., A. Forndran, and K. Dunkers. 1990. "Demonstration of in-receiving water storage of combined sewer overflows in a marine/estuarine environment by the flow balance method, drainage systems, and runoff reduction." Proceedings of the Fifth International Conference on Urban Storm Drainage, Osaka, Japan. Ed. Y. Iwasa and T. Sueishi, vol. 2: 759–64.

Field, R., R. Pitt, M. Brown, and T. P. O'Connor. 1995. "Combined sewer overflow control using storage in seawater." *Water Research.* Elsevier Science Ltd. 30(6): 1505–14.

Field R., and T. P. O'Connor. 1996. "Swirl technology: Enhancement of design, evaluation, and application." *Journal of Environmental Engineering,* American Society of Civil Engineering 122(8).

———. "Optimization of CSO storage and treatment systems." *Journal of Environmental Engineering,* American Society of Civil Engineering 123(3): 269.

Forndran, A., F. Field, K. Dunkers, and D. Moran. 1991. "Balancing flow for CSO abatement." *Water Environment and Technology,* Water Environment Federation 3(5): 54–58.

Gavle, D. R., and D. G. Mitchell. 1996. "Innovative and economical sanitary sewer overflow treatment using fine screens and chlorination." Seminar Publication: National Conference on Sanitary Sewer Overflows. Office of Research and Development, Washington, D.C. EPA/625/R-96/007, pp. 511–19.

Grande, N., and G. Novac. 1995. "New innovation for cleaning sediments in sewers and in storage tanks," WEF Specialty Conference-Sewers of the Future, Houston, Texas.

Gupta, M. K., D. G. Mason, M. J. Clark, T. L. Meinholz, C. A. Hansen, and A. Geinopolos. 1977. "Screening/flotation treatment of combined sewer overflows," vol. 1, *Bench scale and pilot plant investigations.* EPA, Cincinnati, OH, Rep. EPA-600/2-77-065 (NTIS PB 272 834).

Heaney, J. P., W. C. Huber, M. A. Medina, M. P. Murphy, S. J. Nix, and S. M. Hasan. 1977. "National evaluation of combined sewer overflows and urban stormwater discharges," vol. 2. Cost assessment and impacts. EPA, Cincinnati, OH, Rep. EPA-600/2-77-064b (NTIS PB 266 005).

Hedges, P., P. Lockley, and J. Martin 1992. "A field study of hydrodynamic separator combined sewer overflow." International Conference of Innovative Technologies in the Domain of Urban Storm Water Drainage, Novatech 92, Lyon, France.

HIL Technology Inc. 1991. Brochure.

Homack, P. H., et al. 1973. "Utilization of trickling filters for dual treatment of dry and wet weather flows." EPA, Washington, D.C., Rep. EPA-670/2-73-071, NTIS PB 231 251.

Kaufman, H. L., and F. H. Lai. 1978. "Conventional and advanced sewer design concepts for dual purpose flood and pollution control—A preliminary case study," Elizabeth, NJ. EPA, Cincinnati, OH, Rep. EPA-600/2-78-90 (NTIS PB 285 663).

Lager, J. A., and W. G. Smith. 1974. "Urban stormwater management and technology: An assessment." EPA, Cincinnati, OH, Rep. EPA-670/2-74-040 (NTIS 240 687).

Lager, J. A., W. G. Smith, and G. Tchobanoglous. 1977a. "Catchbasin technology overview and assessment." EPA, Cincinnati, OH, Rep. EPA0-600/2-77-051 (NTIS PB270 092).

Lager, J. A., W. G. Smith, W. G. Lynard, R. M. Finn, and E. J. Finnemore. 1977b. "Urban stormwater management and technology: Update and users' guide." EPA, Cincinnati, OH, Rep. EPA-600/8-77-014 (NTIS PB 275 654).

Leiser, C. P. 1974. "Computer management of a combined sewer system." EPA ORD, Washington, D.C., Rep. EPA-670/2-74-022, NTIS PB 235 717.

Lygren, E., and T. Damhaug. 1986. "The swirl concentrator as an urban runoff treatment device." Proceedings of the NATO Advanced Research Workshop on Urban Runoff Pollution, August 26–30, 1985, Montpellier, France, pp. 713–24.

Maher, M. B. 1974. "Microstraining and disinfection of combined sewer overflows—Phase III." EPA, Cincinnati, OH, Rep. EPA-670/2-74-049 (NTIS PB 235 771).

Mahida, V. U., and F. J. DeDecker. 1975. "Multi-purpose combined sewer overflow treatment facility, Mount Clemens, Michigan." EPA, Cincinnati, OH, Rep. EPA-670/2-75-010 (NTIS PB 242 914).

Markland, E., and H. Brombach. 1987. "Air-regulated saddle siphon for stormwater overflow." Fourth International Conference on Urban Storm Drainage, Lausanne, Switzerland.

Meinholz, T. L., D. A. Gruber, R. A. Race, C. A. Hansen, J. H. Moser, and M. J. Clark. 1979. "Screening/flotation treatment of combined sewer overflows—Vol. II; Full-scale operation, Racine, Wisconsin." EPA, Cincinnati, OH, Rep. EPA-600/2-79-106a (NTIS PB80-130693).

Michelbach, S., and C. Wöhrle. 1993. "Settleable solids in a combined sewer system, settling characteristics, heavy metals, efficiency of storm water tanks." *Water Science Technology,* 27 (5–6): 153–64.

Moffa, P. E., E. C. Tiffit, S. L. Richardson, and J. E. Smith. 1975. "Bench-scale high-rate disinfection of combined sewer overflows with chlorine and chlorine dioxide." EPA, Cincinnati, OH, Rep. EPA-670/2-75-021 (NTIS PB 242 296).

Morrissey, S. P., and D. R. F. Harleman. 1992. "Retrofitting conventional primary treatment plants for chemically enhanced primary treatment in the U.S.A." Proceedings of 5th Gothenburg Symposium, Nice, France. *Chemical Water and Wastewater Treatment II.* Springer-Verlag Berlin Heidelberg, pp. 401–16.

Muhs, P. 1995. "Lincoln Park combined sewer overflow treatment operational study." Sanitary District of Decatur, Ill.

Novac, G., N. Grande, and W. C. Pisano. 1996. "Comparison of vortex and hydroslide regulators used to regulate flow in interceptors and sewers." WEF Wet Weather Speciality Conference, Quebec City, Quebec, Canada.

O'Brien and Gere Engineers, Inc. 1992. "CSO abatement program: Segment I—Performance evaluation." Water and Sewer Utility Administration, Washington, D.C.

Parente, M., K. Stevens, and C. Eiker. 1995. "Evaluation of new technology in the flushing of detention facilities." WEF 68th Annual Conference, Miami, Florida.

Parks, J. W., et al. 1974. "An evaluation of three combined sewer overflow treatment alternatives (Shelbyville, IL)." EPA, Rep. EPA-670/2-74-079 (NTIS PB 239 115).

Pisano, W. 1990. "Physical modeling experimental results for design of CSO and stormwater vortex solids separator treatment devices," WEF Annual Conference, Washington, D.C.

Pisano, W. C., G. L. Aronson, C. S. Queiroz, F. C. Blanc, and J. C. O'Shaughnessy. 1979. "Dry-weather deposition and flushing for combined sewer overflow pollution control." Rep. EPA, Cincinnati, OH, Rep. EPA-600/2-79-133 (NTIS PB 80-118 524).

Pisano, W. C., D. J. Connick, and G. L. Aronson. 1984. "Swirl and helical bend regulator/concentrator for storm and combined sewer overflow control." EPA, Edison, NJ, Rep. EPA-600/2-84/151 (NTIS PB 85-102 523).

Sonnen, M. 1977. "Abatement of deposition and scour in sewers." EPA, Cincinnati, OH, Rep. EPA/600/2-77-212 (NTIS PB 276 585).

Springfield Sanitary District. 1970. "Retention basin control of combined sewer overflows." EPA, Water Quality Office, Rep. 11023-0870 (NTIS PB 200 828).

Standard methods for the examination of water and wastewater. 1992. Published jointly by the American Public Health Association, American Water Works Association, and Water Environment Federation, 18th ed., section 2540F.

Sullivan, R. H., M. M. Cohn, J. P. Coombes, and B. S. Smisson. 1972. "The swirl concentrator as a combined sewer overflow regulator facility." EPA, Edison, NJ, Rep. EPA-R2-72-008 (NTIS PB 214 687).

Sullivan, R. H., M. M. Cohn, J. E. Ure, F. E. Parkinson, and G. Galiana. 1974. "Relationship between diameter and height for design of a swirl concentrator as a combined sewer over-flow regulator." EPA, Edison, NJ, Rep. EPA-670/2-74-039 (NTIS PB 234 646).

Sullivan, R. H., et al. 1982. "Assessment of sewer sealants." EPA, ORD, Cincinnati, OH, Rep. EPA 600/8-82-012 (NTIS PB 82 262 718).

Sullivan, R. H., J. E. Ure, F. Parkinson, and P. Zielinski. 1982. "Design manual—Swirl and helical bend pollution control devices." EPA, Edison, NJ, Rep. EPA-600/8-82/013 (NTIS PB 82-266 172).

Tchobanoglous, G., and F. L. Burton. 1991. *Wastewater engineering: Treatment, disposal, and reuse.* Metcalf and Eddy, Inc. 3rd ed. McGraw-Hill, Inc., New York.

URS Consultants, Inc. 1990. Hydrodynamic Solids Separation Evaluation. Wards Island, New York, NY.

Urbonas, B., and P. Stahre. 1993. *Stormwater: Best management practices and detention.* Englewood Cliffs, NJ: Prentice Hall.

Walker, D., J. Golden, D. Bingham, and E. Driscoll. 1993. "Manual: Combined sewer overflow control." EPA, Cincinnati, OH, Rep. No. EPA/625/R-93/007 (NTIS PB 93-144 649).

Watt, T. R., R. G. Skrentner, and A. C. Davanzo. 1975. "Sewerage system monitoring and remote control." EPA, Cincinnati, OH, Rep. EPA-600/2-75-020 (NTIS PB 242 107).

Welborn, H. L. 1974. "Surge facility for wet and dry weather flow control." EPA, Cincinnati, OH, Rep. EPA-670/2-74-075 (NTIS PB 238 905).

Welsh, F. L., and D. J. Stucky. 1974. "Combined sewer overflow treatment by the rotating biological contactor process." EPA, Cincinnati, OH, Rep. EPA-670/2-74-050 (NTIS PB 231 892).

Western Company. 1969. "Polymers for sewer flow control." Federal Water Pollution Control Administration, Rep. 11020DIG08/69 (NTIS PB 185 951).

White, G. C. 1986. *Handbook of chlorination.* 2d ed. New York: Van Nostrand Reinhold.

Wong, T. H. F., and R. M. Wooton. 1995. "An innovative gross pollutant trap for stormwater treatment." Proceedings of 2nd International Symposium on Urban Stormwater Management, Melbourne, Australia, pp. 407–12.

Bibliography

American Public Works Association. 1970. "Control of infiltration and inflow into sewer systems." Federal Water Pollution Control Administration, Rep. 11022EFF 12/70 (NTIS PB 200 827).

———. 1971. "Prevention and correction of excessive infiltration and inflow into sewer systems—A manual of practice." EPA, Washington, D.C., Rep. 11022EFF01/71 (NTIS PB 203 208).

Averill, D., et al. 1995. "Pilot testing of high-rate treatment methods for CSO control, new and emerging environmental technologies and products for wastewater treatment and stormwater collection." Water Environment Federation, Toronto.

Boner, M. C., S. R. Harper, G. C. Patrick, D. R. Ghosh, and W. L. Cairns. 1994. "Optimization of vortex separator removal efficiencies for CSO treatment." Project 92-TCR-2, Preliminary findings, Water Environment Research Foundation, Alexandria, VA.

Brombach, H. 1984. "Vortex flow controllers in sanitary engineering." Trans. ASME Jour Dynamic Systems Measurements and Control, vol. 106.

———. 1989. "Equipment and instrumentation for CSO control." Engineering Foundation Conference: Urban Stormwater Quality Enhancement-Source Control, Retrofitting and Combined Sewer Technology, Davos, Switzerland.

Brombach, H., C. Xanthopoulos, H. H. Hahn, and W. C. Pisano. 1993. "Experience with vortex separators for combined sewer overflow control." *Water Science Technology,* 27(5–6): 93–104.

Carcich, I. G., et al. 1972. "A pressure sewer system demonstration." EPA, Washington, D.C., Rep. EPA-R2-72-091 (NTIS PB 214 409).

Central Engineering Laboratories, FMC Corp. 1972. "A flushing system for combined sewer cleansing." EPA, Washington, D.C., Rep 11020DNO03/72, (NTIS PB 210 858). Santa Clara, CA.

Cesareo, D. J., and R. Field. 1975. "Infiltration-inflow analysis." *Journal of the Environmental Engineering Division,* ASCE, vol. 101, no. EE5, Proc. Paper 11645 (October 1975): 775–85.

Connick, D. J., et al. 1981. "Evaluation of a treatment lagoon for combined sewer overflow." EPA, Rep EPA-600/2-81-196 (NTIS PB 82-105214).

Cronk, G. E. 1972. "Ground water infiltration and internal sealings of sanitary sewers, Montgomery County, OH." EPA, Washington, D.C., Rep. 11020DHQ 06/72 (NTIS PB 212 267).

Deleporte, C., et al. 1995. "Optimized Lamellae settling of urban stormwater waste." *Water Science and Technology,* 32(1).

Drake, J., and C. Hunsinger. 1990. "CSO treatment utilizing revolutionary swirl concentrator. Sanitary District of Decatur, Ill." Central States WPCA Annual Conference, Wisconsin.

Field, R. 1980. "EPA research in urban stormwater pollution control." *Journal of the Hydraulics Division,* ASCE, vol. 106, no. HY5, Proc. Paper 15389 (May 1980): 819–35.

———. 1986. "State-of-the-art update on combined sewer overflow control." CRC Press Inc., Cat. No. 1367, *Critical Reviews in Environmental Control* 16(2).

Foster, N. S., and R. H. Sullivan. 1977. "Sewer infiltration and inflow control product and equipment guide." EPA, Cincinnati, OH, Rep. EPA 600/2-77-017c (NTIS PB 272 680).

Giordano, J. J. 1978. "The construction, technical evaluation, and frictional determination of an aluminum storm sewer system." EPA, Cincinnati, OH, Rep. EPA-600/2-78-025 (NTIS PB 283 458).

Guibelin, E., et al. 1994. "The Actiflow process: A highly compact and efficient process to prevent water pollution by stormwater flows." IAWQ, Water Quality Control Conference, Budapest.

Hoffman and Fiske. 1969. "Reduction of groundwater infiltration into sewers by zone pumping at Meridian, Idaho." Federal Water Pollution Control Administration, Rep. 11020-06/69 (NTIS PB 187 868).

Hubner, M., and W. Geiger. 1995. "Review of hydrodynamic separator-regulator efficiences for practical applications." *Water Science and Technology,* IAWQ, 32(1).

Innerfeld, H., A. Forndran, D. D. Ruggiero, and T. J. Hartman. 1979. "Dual process high-rate filtration of raw sanitary sewage and combined sewer overflows." EPA, Cincinnati, OH, Rep. EPA-600/2-79-015 (NTIS PB 296 626/AS).

Kaufman, H. L., and F. Lai. 1980. "Review of alternatives for evaluation of sewer flushing-Dorchester Area-Boston." EPA, Cincinnati, OH, Rep. EPA-600/2-80-118 (NTIS PB81-221 872).

Mayer, J. K., et al. 1972. "Sewer bedding and infiltration in the Gulf Coast area." EPA, Washington, D.C., Rep. 11022DEI05/72 (NTIS PB 211 282).

Metropolitan Sewer Board. 1971. "Dispatching systems for control of combined sewer losses." EPA, Washington, D.C., Rep. 11020FAQ03/71 (NTIS PB 203 678).

O'Shea, M. L., and R. Field. 1992. "An evaluation of bacterial standards and disinfection practices used for the assessment and treatment of stormwater." *Advances in Applied Microbiology* 37: 21–40.

Olson, M. R. 1985. "Insituform and other sewer rehabilitation techniques." EPA, Cincinnati, OH, Rep. EPA-600/2-85/132 (NTIS PB 86-130 192).

Pisano, W. 1994. "Sieving treatment for CSO floatables control." WEF Specialty Conference Global Perspective for Reducing CSOs, Louisville, Ky.

Pisano, W. C., and C. S. Queiroz. 1977. "Procedures for estimating dry-weather pollutant deposition in sewerage systems." EPA, Cincinnati, OH, Rep. EPA-600/2-77-120 (NTIS PB270 695).

Pisano, W. C., et al. 1983. "Value of flow calibration for decision making in infiltration/inflow studies." EPA, Cincinnati, OH, Rep. EPA-600/2-83/087 (NTIS PB 83-259 739).

Pisano, W., and H. Brombach. 1993. "Historical perspective: Use of vortex flow throttles as flow controllers in sewerage systems." AWPCS/WEF Specialty Collection Systems Operation Conference, Tuscon, Arizona.

———. 1995. "New European innovations for combined sewer overflow control," in *WEF new and emerging technologies for collection and treatment of wastewater: A global perspective,* Toronto, Canada.

Shelly, G. J., P. B. Stone, and A. J. Cullen. 1981. "Field evaluation of a swirl degritter at Tamworth, New South Wales, Australia." EPA, Edison, NJ, Rep. EPA-600/2-81-063 (NTIS PB 81-187 247).

Shuckrow, A. J., et al. 1973. "Physical-chemical treatment of combined and municipal sewage." EPA, Washington, D.C., Rep. EPA-R2-73-149 (NTIS PB 219 668).

Steketee, C. H. 1985. "Demonstration of service lateral testing and rehabilitation techniques." EPA, Cincinnati, OH, Rep. EPA-600/2-85/131 (NTIS PB 86-135 647).

Sullivan, R. H., et al. 1977. "Economic analysis, root control as related to infiltration/inflow: Volume I—Final report." EPA, Cincinnati, OH, Rep. EPA-600/2-77-017a, (NTIS PB 280 738).

———. 1977. "Sewer system evaluation rehabilitation and new construction—A manual of practice." EPA, Cincinnati, OH, Rep. EPA 600/2-77-017d, (NTIS PB 279 248).

Sullivan, R. H., J. E. Ure, and P. Zielinski. 1977. "Field prototype demonstration of the swirl degritter." EPA, Edison, NJ, Rep. No. EPA-600/2-77-185 (NTIS PB 272 668).

Sullivan, R. H., and J. W. Ewing. 1985. "Selected topics related to infiltration and inflow in sewer systems." EPA, Cincinnati, OH, Rep. EPA-600/2-85/020 (NTIS PB 85-102 523).

Chapter 5

Cost-Effective Analysis

- PETER E. MOFFA, P.E.
  Principal
  Moffa and Associates, Consulting Engineers
  Syracuse, New York

GENERAL

In simple terms, a cost-effective solution is one in which you get the most for your money. When costs are plotted against benefits, the break, or the "knee" of the curve, is that point where additional benefits result in more-rapidly-rising unit costs, or the period beyond which there are diminishing returns, as illustrated by Figure 5–1. The stormwater and CSO problem presents a high degree of variability in terms of rate of flow, volume, pollutants discharged, and water-quality impact. This has undoubtedly contributed to the indecision by regulatory agencies to come to grips with rules governing CSO abatement. Nevertheless, it was not until 1989, when the EPA published the National CSO Control Strategy, that states developed their own requirements for their respective communities to follow. All but a few municipalities embarked upon their Long Term Control Plan (LTCP) before states issued their own requirements. Even to this day there exists a certain amount of confusion as to how to develop an LTCP and all the elements that comprise such a plan. In consideration of communities that are still grappling with their LTCP, these case studies emphasizing cost benefit are presented in this chapter. This approach becomes particularly important in light of the dwindling nature of federal and state monies and the greater reliance on local monies. Local authorities will require more information on the return on a major investment to improve water quality before being willing to proceed. Also, there may be some tendency to challenge the legality of water-quality standards unless reasons for doing so are made clear.

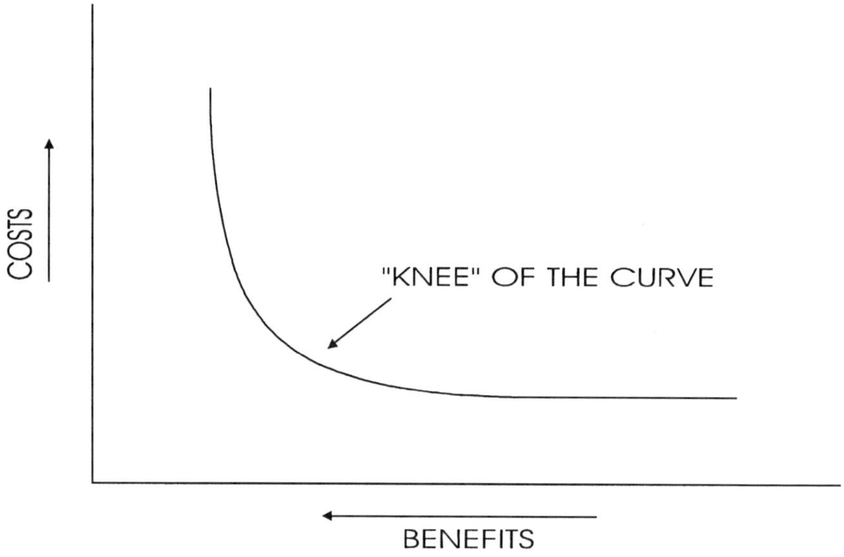

Figure 5-1. Cost-benefit relationship.

Benefits

There are four general methods for estimating benefits of water-quality management projects (EPA-625/4-79-013):

1. The first is the simplest. It assumes that benefits will automatically be realized from the proposed activity and that they are so large that they justify the allocation of required resources. It completely avoids the benefit measurement problem. An example of this method is the federal requirement for secondary treatment at publicly owned treatment works.

2. In the second method, certain water-quality standards are associated with the protection of water uses. This is generally the case, although the connection may not always be scientifically defensible. Once standards to protect desired beneficial uses have been established, the comparison of projected water quality with the standards provides an estimate of benefit.

3. Estimates of increased water-use potential are developed by translating improvements in water quality into numerical increases in usable length of beaches, swimming days, fishing days, or waterfront property values, to name a few possibilities. This is in effect a refinement of the water-quality standards approach, allowing benefits to be estimated directly.

4. The best known method is the classical cost-benefit analysis. In this method, each anticipated water usage has a dollar value assigned as a measure of unit worth or willingness to pay. The actual project usages are multiplied by the corresponding unit values, and the sum is compared directly to total project costs.

The first approach is unacceptable for CSO control planning because the benefits are not obvious; they are highly contingent on both stormwater volumes and the receiving water. The second approach, in and of itself, is difficult at best. It poses a near-impossible situation for regulatory agencies to address, particularly during the storm when the overflow and receiving waters are both changing rapidly. The second approach combined with the third would appear to offer the most effective means of obtaining approvals and funding from all levels—that is, relating mandated water-quality requirements to lost resources and real benefits of recovery. The fourth approach involves too much subjectivity and could result in lengthy dialogue over the unit worth of various usage on both the local and regulatory levels.

Design Conditions

Once data are gathered and modeling is performed to characterize discharges from the CSO system and identify water-quality impacts, a basic decision becomes necessary: What combination of storm- and receiving-water conditions should be selected for abatement? This in turn should depend on the water-quality objective or target parameter (i.e., aesthetics [solids and floatables], dissolved oxygen, bacteria, nutrients, etc.).

The dry-weather approach of selecting the "minimum seven-day consecutive flow over a 10-year return period" (MA7CD10) for the receiving stream allows a once-in-ten-year violation for a fixed discharge quantity. However, a second and more important consideration when dealing with CSO abatement is the rainfall that results in a highly variable discharge as opposed to dry-weather conditions. Rainfall is characterized by a combination of average storm intensity and duration, as illustrated by Figure 5–2. The introduction of this second variable requires a closer inspection of when each of the two conditions—the selected rainfall event and the minimum receiving water flow—will occur. Otherwise, the probability of these two conditions jointly occurring can easily exceed 10 years. Thus, it becomes necessary to review rainfall records in combination with receiving-water flow records before selecting design conditions. A typical plot of rainfall per storm for each of the months for Syracuse, New York, best illustrates that the more intense rainfall events occur during the summer months (see Figure 5–3). Assuming recreation such as boating and water skiing could take place as

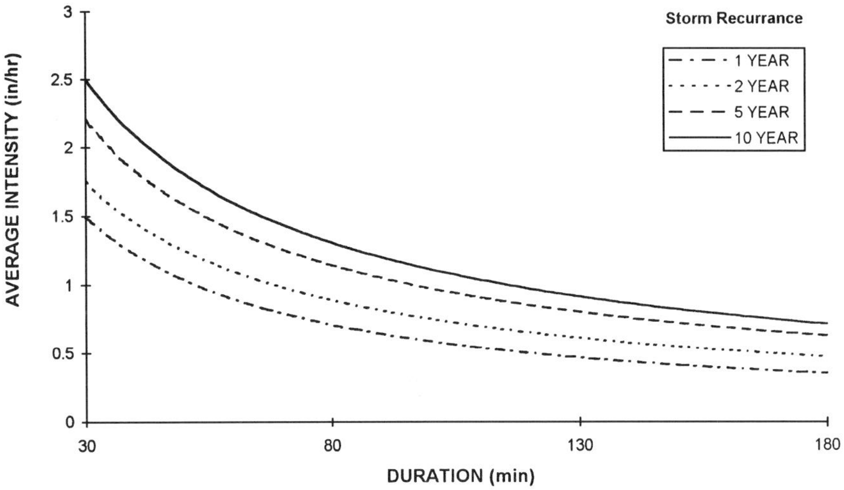

Figure 5-2. Intensity-duration curves.

early as May and extend as late as October 31, it would be reasonable to consider receiving-water flows for this six-month period.

A third consideration that has been introduced as a result of the EPA regulation of CSOs is the capacity of the combined sewer system. The draft policy states that "all CSOs must be brought into compliance with technology-based requirements and state water-quality based requirements. The agency expects that this can be achieved using a combination of CSO control measures" (AMSA 1988). This implies control and treatment of the largest storm that could be conveyed by a sewer system. Thus, the capacity of trunk and interceptor sewers becomes the limiting factor in determining the frequency of storms that will have to be controlled and treated. More restrictive systems that cannot convey higher rates of flow will require smaller abatement facilities. Rates of flow that exceed the system capacity will result in street flooding, which would have to be either tolerated or treated as a separate stormwater flow. Thus, a cost-effective methodology for CSOs must consider combinations of (1) rainfall intensity, (2) receiving-water flow, and (3) system capacity. Cost-benefit relationships should be developed for each of the receiving-water-quality parameters of concern. Generally, the periods of treatment for water-quality pollutants from CSOs can be classified as follows:

Parameter	Period of Treatment
Solids and Floatables	All discharge events
Dissolved Oxygen	Combinations of rainfall events and receiving-water flows equal to or less than a 10-year recurrence interval

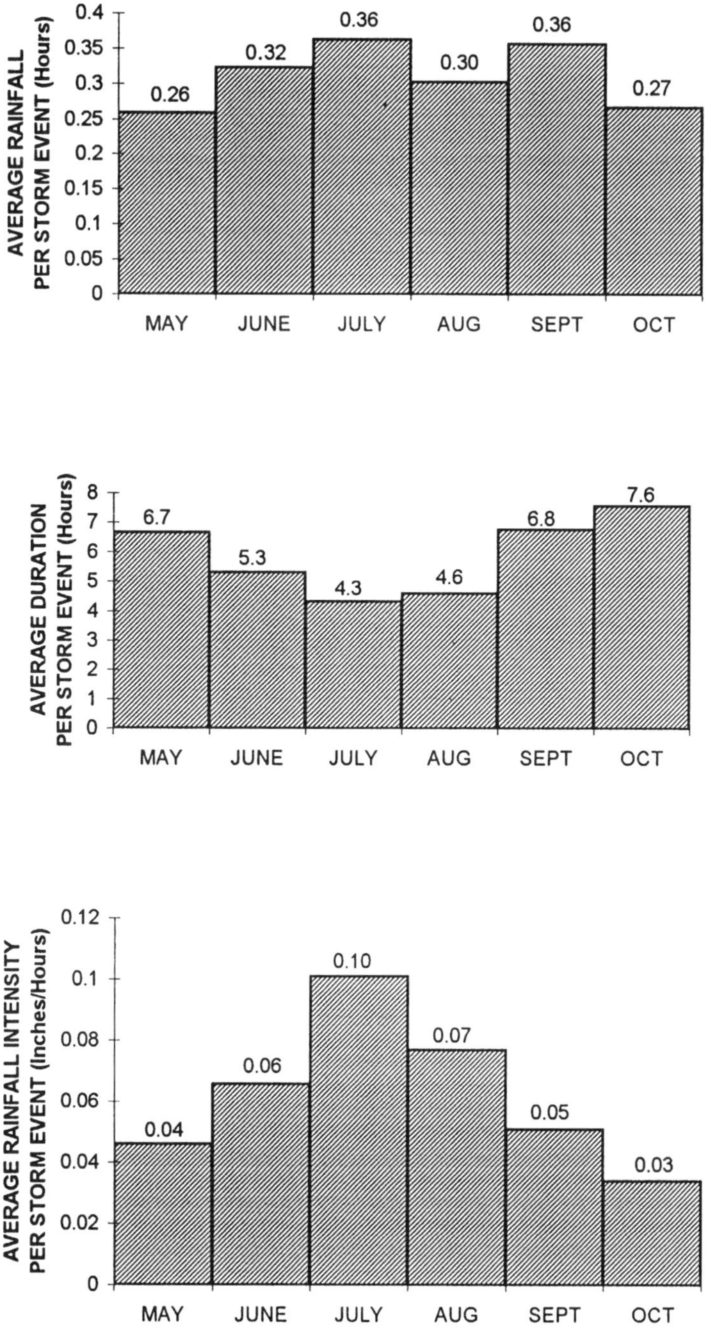

Figure 5-3. Rainfall characteristics for Syracuse, New York.

Bacteria	Recreational season (e.g., May–October)
Toxics	All of the time for specific tributary areas, as determined through sampling
Nutrients	Periods of the year as determined from relative significance of CSOs as compared to other nutrient sources

METHODOLOGY

This chapter outlines a general methodology for determining a cost-effective solution citing case histories as appropriate. The general approach for cost-effective analysis for CSO abatement is illustrated by Figure 5–4.

Subfigure A illustrates the relationship between total CSO volume discharged from a sewer system and accumulative peak rate versus the storm size in terms of frequency of return. Generally, frequency of return is based on a constant duration such as two hours and a varying rainfall intensity. The SWMM can be utilized to develop total volume discharged, which represents the total of all overflows irrespective of their points of discharge. This is valid if one desires to develop a uniform set of design conditions throughout a sewer system. In large systems with a wide range of types of receiving waters (e.g., small streams and large bays), it may be desirable to develop this relationship for each specific water body. Such a curve can also be developed for accumulative peak rate, which is a summation of the peak rates of all overflows at a given frequency storm, irrespective of the timing of the peak. This parameter is an indication of the peak carrying capacity of the system and complements the plot of total CSO volume. These data can then be plotted to yield a curve as illustrated by Subfigure A, which can be used to designate the "selected storm" storm condition. At some storm frequency, the system can no longer accept any more stormwater flow, resulting in street ponding and either subsequent discharge to the system or evaporation or infiltration into the ground. In the case of Onondaga County, the sewer system shows a significant break beyond a storm frequency of one year, as illustrated by Figure 5–5. Historical records have verified that combined sewer systems were designed to accommodate a storm with a frequency return of approximately one year.

Subfigure B illustrates the relationship between receiving-water flow and the extent of violation at the selected design storm (e.g., one year frequency of return). This relationship would have to be plotted for each of the water-quality parameters of concern. This particular curve is more illustrative of the relationship for bacteria where the degree or extent of violation increases as the dilution from water flow decreases. At some point in the decision-making process, a

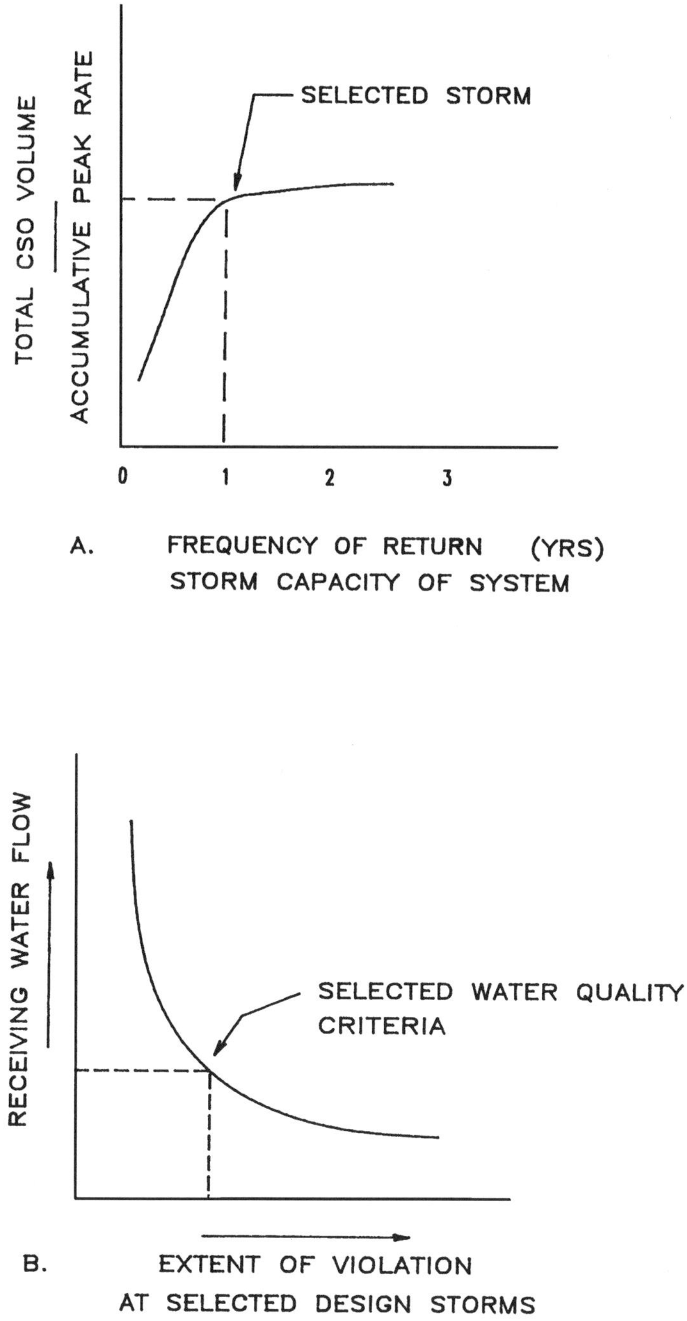

Figure 5-4. General approach for cost-effective analysis for CSO abatement.

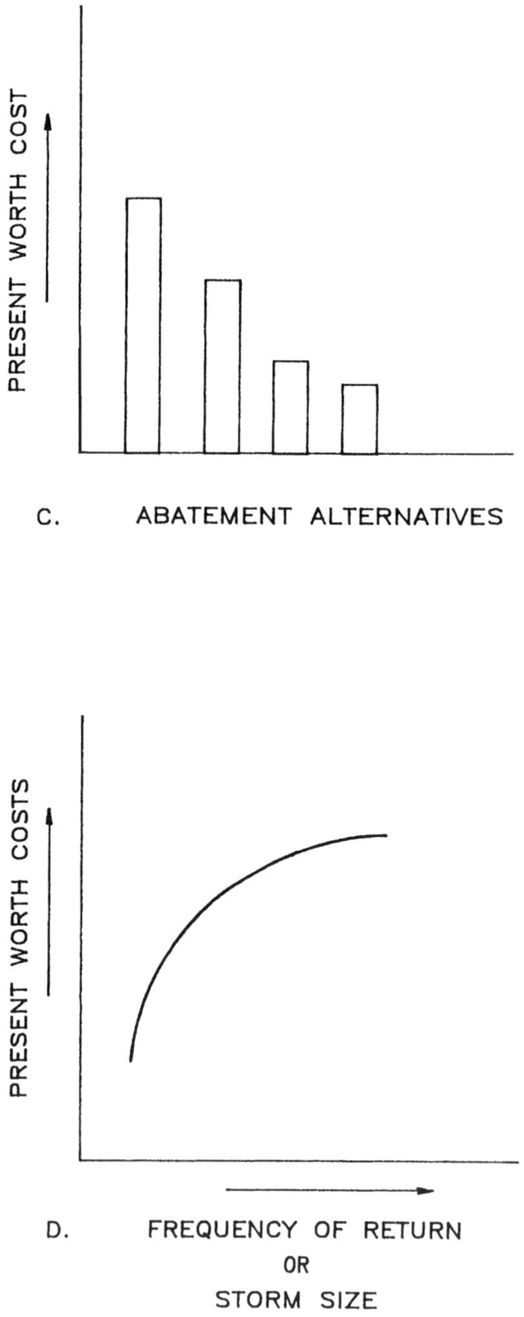

Figure 5–4. Continued.

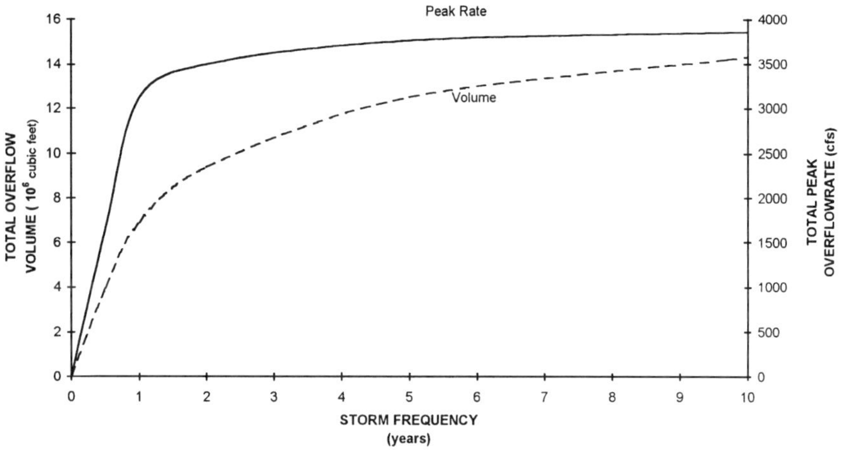

Figure 5–5. Storm frequency vs. peak rate and volume for Syracuse, New York.

receiving-water flow would have to be selected so as to be reasonable when combined with the probability of the particular frequency of storm that is chosen (see "Critical Design Conditions" in chapter 3). One possibility of a stream flow that could be selected is that for the dry months of July, August, and September, when the highest intensity storms might be expected, as illustrated in Figure 5–3. Upon performing this evaluation for each of the water-quality parameters in question, the required removals can then be determined for each parameter.

Subfigure C, given the removals required, can then be developed on the basis of specific abatement technology. Various abatement alternatives that would accomplish the same reductions can be evaluated with respect to present-worth costs. Costs have been selected merely as a comparative term. In many cases, the Y axis may represent a combination of regulatory, institutional, and local requirements in addition to costs.

Subfigure D, given the selected abatement alternative, is a plot of the variation of different-sized facilities for that alternative, as required to treat different-sized storms against the relative costs. In this way, a final evaluation of the most cost-effective-sized facility can be determined.

Case Histories

The following are excerpts taken from three CSO studies that developed cost-benefit relationships:

Augusta, Maine

The combined sewered area of Augusta, Maine, comprises approximately 700 acres with 36 diversion structures and 20 active CSOs that discharge either directly to or into tributaries of the Kennebec River, which serves as an excellent fishery as well as a heavily-used recreation area. In December, 1993, Whitman and Howard (now Earth Tech) completed a CSO facility plan that evaluated eleven alternatives for storm intensities ranging from ½-year return frequency to a five-year return frequency storm. Storm frequencies that were investigated were based upon the system capacity curve shown in Figure 5–6. Due to the large size of the Kennebec River, the primary water quality was that of bacteria.

Figure 5–7 illustrates maximum *E. coli* concentrations in the Kennebec River at a minimal flow of 1,000 cfs for different recurrence interval storms. As shown, the one-year recurrence interval represents the knee of the curve. Consequently, a series of abatement alternatives was evaluated for the one-year storm. Present-worth costs for several prime alternatives are illustrated in Figure 5–8. The least costly alternative is shown as Alternative 3 and involves two vortex sites; however, the alternative selected was the next costly, which included a single vortex at the existing WWTF. The alternative was selected due to nonmonetary factors such as facility siting, the desire to avoid satellite facilities, and flexibility for future expansion or upgrades.

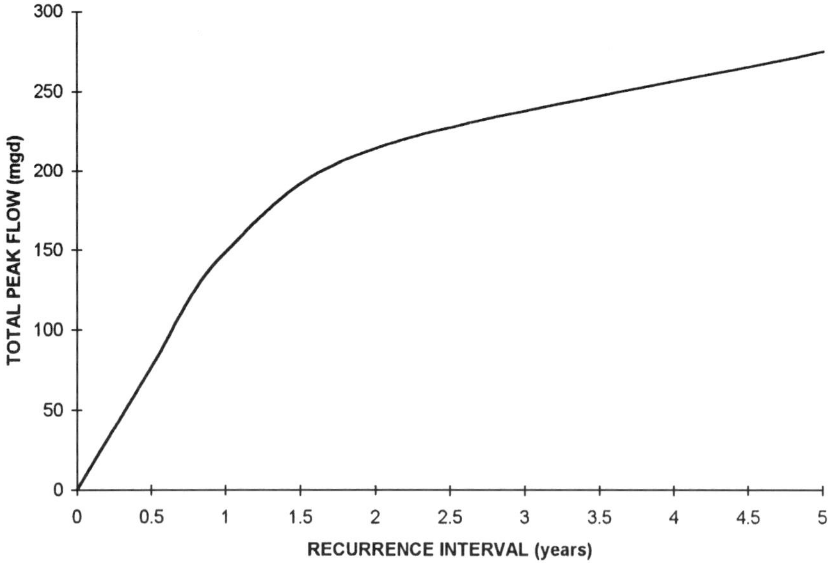

Figure 5–6. Recurrence interval vs. total peak flow, Augusta, Maine.

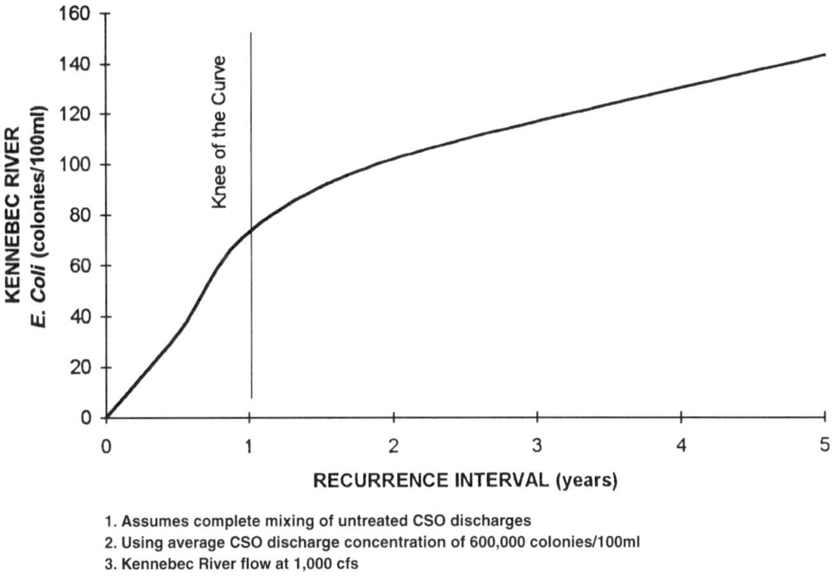

Figure 5-7. Recurrence interval vs. *E. coli* count, Augusta, Maine.

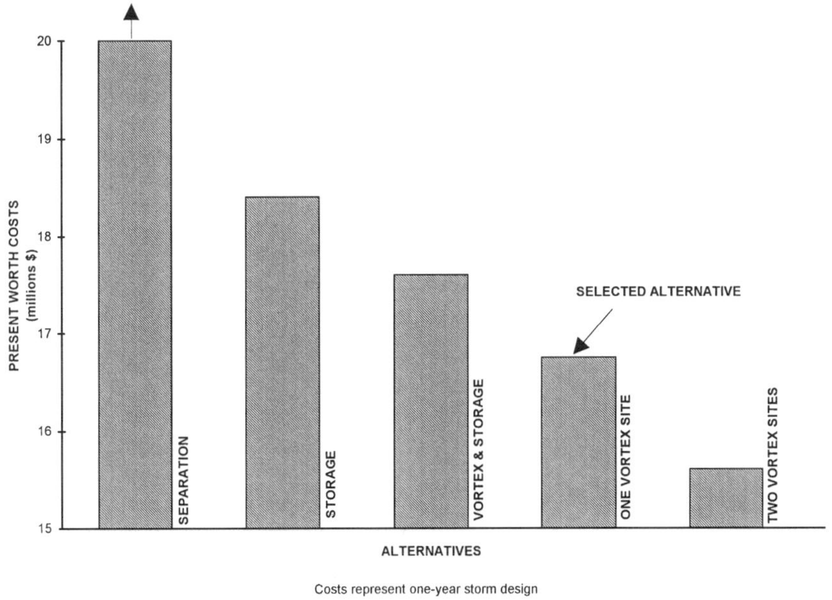

Figure 5-8. Present worth costs for CSO abatement alternatives, Augusta, Maine.

The selected solution involves constructing consolidation conduits, two CSO pumping stations, and a single vortex treatment facility at the site of the WWTF, which is being expanded to include a CSO-related bypass to treat roughly 44% of Augusta's current annual volume. This solution has been approved by the EPA and the State of Maine DEP. Given the selected treatment alternative, costs were generated for different size storms as illustrated in Figure 5–9.

Syracuse, New York

Syracuse is a city of 164,000 people located at the foothills of the Appalachian plateau in Onondaga County, New York. Approximately 40% of the city's 16,600 acres is served by a combined sewer system. Each of the 66 combined sewer overflows (CSO) discharge into one of three receiving streams that are tributary to Onondaga Lake as shown on Figure 5–10. Onondaga Lake has become a matter of national interest as one of the most polluted lakes in the United States. This urban lake, approximately 4.5 square miles in area, and its drainage area lie wholly within Onondaga County and serves as a good opportunity to assess CSO impacts within the context of an entire watershed.

After more than 25 years of evaluating the CSO problem involving demonstrating various abatement technologies, characterizing the sewer system and re-

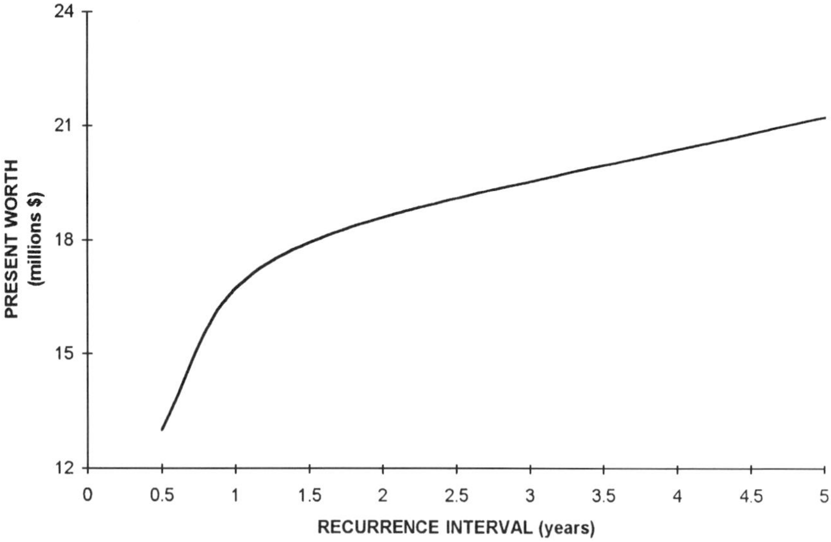

Figure 5–9. Recurrence interval vs. present worth for select abatement alternative, Augusta, Maine.

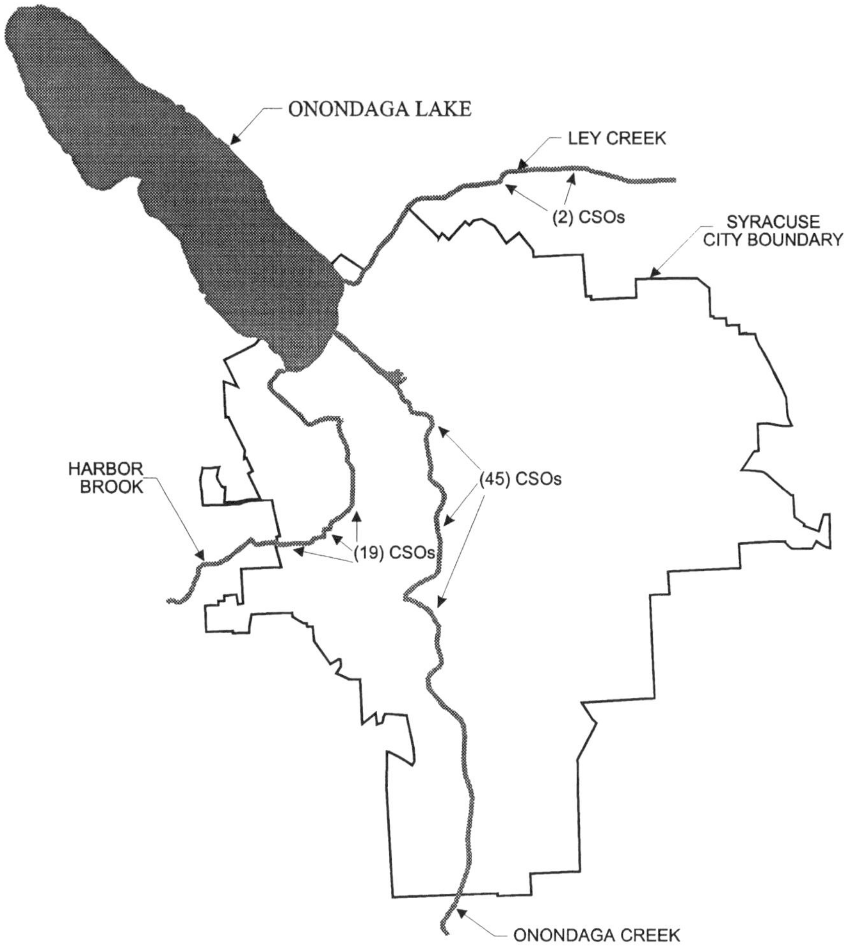

Figure 5–10. Combined sewer overflow tributary area, Onondaga County, New York.

ceiving waters, implementing a BMP program, and performing extensive cost/benefit analyses, a comprehensive CSO abatement program for the City of Syracuse, estimated at over $140 million, is underway. In 1987 the county voluntarily undertook an update of a CSO facilities plan (1979). A year later the county entered into a consent order initiated by the Atlantic States Legal Foundation and enjoined by the NYS Department of Environmental Conservation. The basis for this legal action was a series of phosphorous violations at the Syracuse

Metropolitan sewage treatment plant; nevertheless, CSO abatement became part of the Municipal Compliance Plan.

As part of the facilities plan, the Storm Water Management Model (SWMM) in conjunction with flow monitoring was utilized to determine CSO discharge flows. Sampling and analyses were employed to determine the quality of CSOs that, combined with the SWMM results, were used to estimate CSO loadings to receiving waters (see chapter 1, Figure 1–4).

Effectiveness of the BMP Program

Flow monitoring, sampling, and analyses were conducted prior to and following a Best Management Practices Program. The results showed more than an 85% reduction in annual pollutant to the receiving waters. The major elements of the BMP program were: sewer cleaning, overflow-structure modifications, and sewer replacement to ensure maximum system storage and conveyance to the Syracuse Metropolitan Sewage Treatment Plant (STP). Table 5–1 illustrates the improvements as a result of a $10 million BMP program conducted in 1982–1984.

System Modeling

The SWMM model was instrumental in determining the conveyance capacity of the system. Once the model was calibrated, different frequency storms were run to determine the accumulative peak flows within the system. A plot of this relationship is shown in Figure 5–5; increasing recurrence intervals are representative of increasing intensities at an arbitrarily selected storm duration of two hours. As can be noted, there is a severe break (knee of the curve) in the system's ability to convey rate of flow at the one-year storm interval. The total-overflow-volume relationship also showed a break, although a less marked one. In general

Table 5–1. CSO Discharge Statistics, Syracuse, New York

ITEM	PRE-BMP	POST-BMP
▪ Number of Events/Year	165	56
▪ Number of Events 6/15–9/15	42	18
▪ Annual CSO Volume (million gallons)	11000*	1657
▪ 1-Year Storm Volume (million gallons)	72	59
▪ 90 Percentile Storm Volume (million gallons)	54	40

*Projected from pre-BMP measurements, which reflects reduced sewer capacities, creek water intrusion, and infiltration

it was decided to consider a one-year storm event as a reasonable basis for estimating the costs of various abatement alternatives.

Sizing and costing the storage facilities involved selecting a one-year storm volume through the use of continuous simulation. The Runoff Block of SWMM in combination with hourly rainfall data was used to project overflow volumes for storms over a 39-year period of record; the one-year volume was selected from this array of volumes. (The importance of using continuous simulation in modeling is described in chapter 2.)

The peak intensity for the one-year storm event as determined through the Keifer-Chu method and the continuous simulation calculations served as the basis for high-rate type facilities and storage facilities, respectively.

Receiving Water Impacts

Water-quality concerns as developed through a Storms Impact Study conducted in the mid-1970s and verified in the early 1990s are summarized in Table 5–2.

A limited watershed effort became necessary to put the CSO component of solids and nutrients in proper perspective. Figure 5–11 illustrates the Onondaga Lake watershed, which lies wholly within Onondaga County. Agriculture comprises the major land activity outside of the City of Syracuse and the urban areas.

Using a watershed model, Hydraulic Simulation Program Fortran (HSPF) in combination with a geographic information system (IDRISI) and SWMM for the urban areas, solids and nutrient loadings were calculated for an average rainfall year (1991). Table 5–3 illustrates the loadings relative to the STP. The net result was that phosphorus, nitrogen, and solids were found to be of less significance than was thought before the watershed estimates were made. Heavy metals were not considered since concentrations as measured in the receiving waters in the 1970s Storms Impact Study were well below levels of concern.

The event-specific parameters as listed in Table 5–2 were evaluated for a one-year, two-hour storm. It was concluded early in the study that floatables were observed following most storms and did not need further evaluation.

Table 5–2. Event-Specific Parameters

Parameter	Nature of Impact
▪ Floatables	▪ Event Specific
▪ Solids	▪ Accumulative
▪ Dissolved Oxygen	▪ Event Specific
▪ Nutrients	▪ Accumulative
▪ Bacteria	▪ Event Specific
▪ Heavy Metals	▪ Accumulative

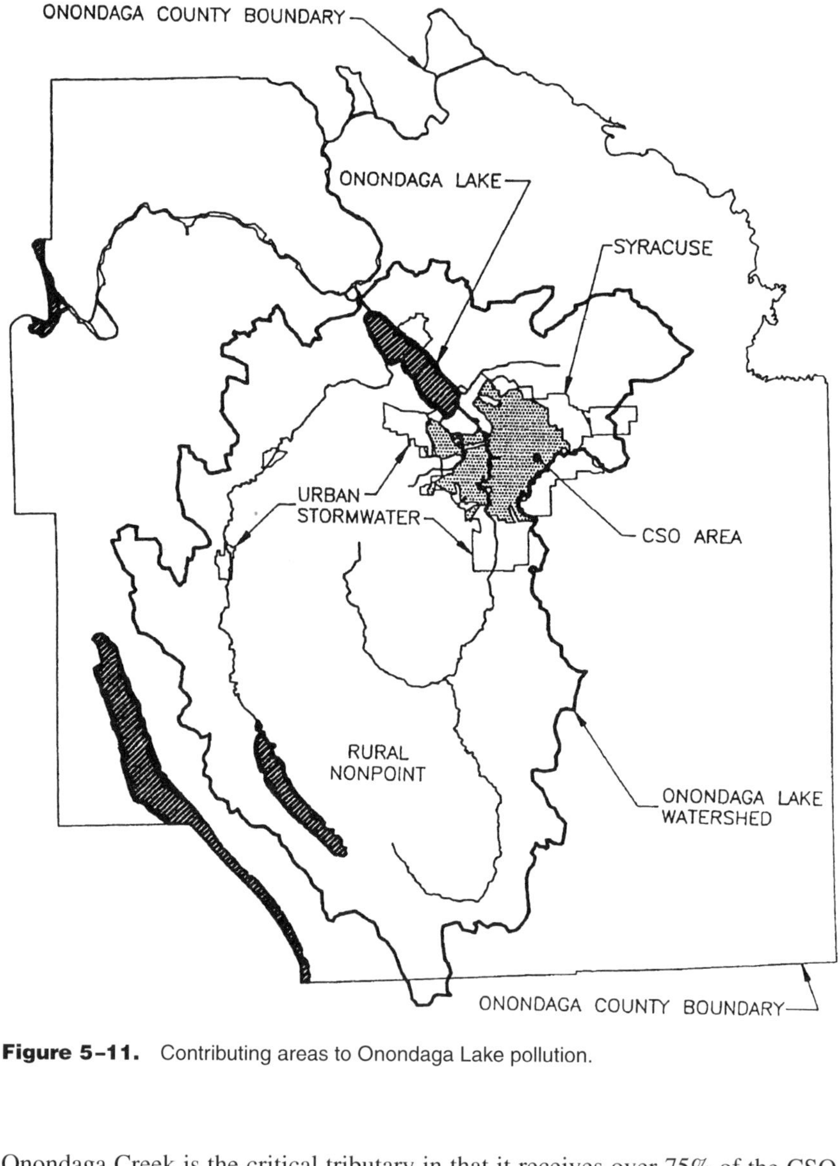

Figure 5–11. Contributing areas to Onondaga Lake pollution.

Onondaga Creek is the critical tributary in that it receives over 75% of the CSO loading and is a potential fish-propagation stream. Therefore, it served as the basis for evaluating dissolved oxygen (D.O.) impact from a one-year CSO event using the 30-day low flow of the period of record. Figure 5–12 shows that the minimum D.O. reaches approximately 5 mg/l, the stream standard for Class B and C in New York State, before the stream joins the lake. Therefore, it was con-

Table 5–3. Loading Summary to Onondaga Lake

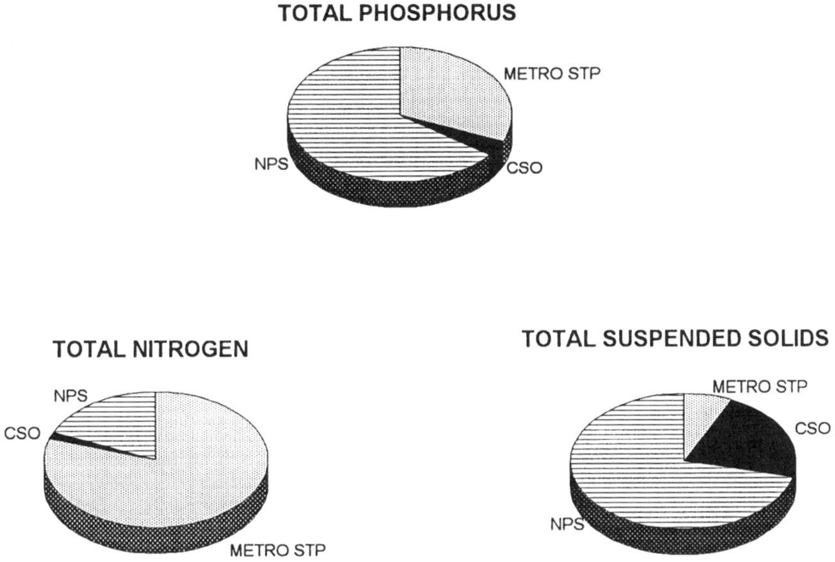

Pollutant Source	Total Phosphorus		Total Nitrogen		Total Suspended Solids	
METRO STP	134,320	31%	5,762,620	80%	1,167	7%
CSO	16,800	4%	125,432	2%	3,739	22%
NPS	275,349	65%	1,313,434	18%	12,107	71%
TOTAL	426,469		7,201,486		17,013	

cluded that D.O. was not a concern. These calculations included SOD measured within the inner harbor.

Bacteria impacts were of primary concern in Onondaga Lake but not in the tributaries that serve as flood-relief channels. Based upon a calibrated bacteria model of the lake and bacterial loadings from all storms in a typical year (1991), the number of violations, defined as exceeding 200 fecal coliform per 100 ml, were calculated for each of the cells within the lake. Cells were governed by distinct spatial and depth boundaries in the lake. Figure 5–13 depicts the number of bacteria violations that occur under existing conditions, reflecting only BMP improvements. The southern area of the lake nearest to the CSO tributaries show the greatest number of annual violations with a decrease in violations progressing toward the northern end of the lake.

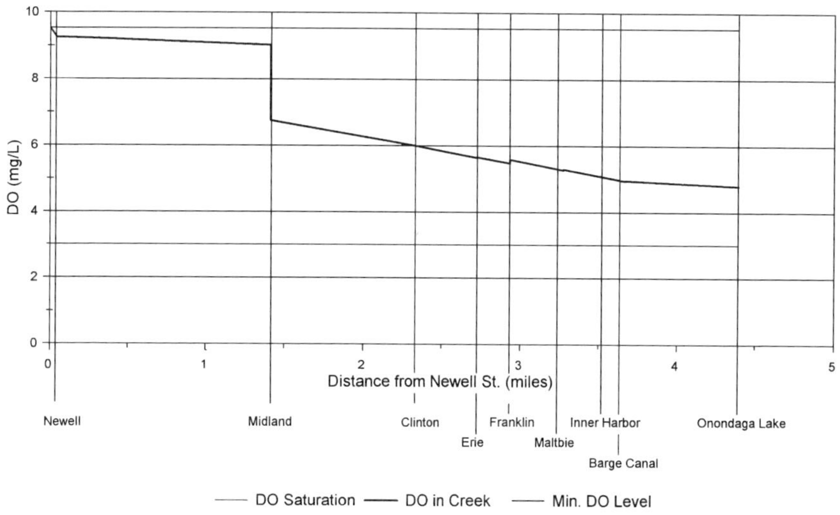

Figure 5–12. Dissolved oxygen concentrations in Onondaga Lake and tributaries.

CSO Abatement

The consent order required that nine specific abatement alternatives be evaluated. The number of abatement alternatives was reduced to five based upon cost comparisons. These were ultimately reduced to the following three:

- Regional Treatment—vortex, high-rate disinfection
- Regional Treatment/Storage—vortex, storage, high-rate disinfection
- Centralized Storage—tunnel storage followed by treatment at STP

Preliminary cost estimates resulted in the two regional alternatives as the focus of further environmental evaluations. The county favored regional treatment on the basis of cost: $268 million. The state preferred to compliment the county alternative with storage for one-half of the one-year storm. This would add $96 million to the cost and bring the total to $364 million. The argument centered on the additional benefit, if any, that additional storage might provide (e.g., phosphorus). The vortex units sized for the instantaneous peak of a one-year event and disinfection tanks sized for five-minute contact of that peak flow effectively provided total storage of smaller storms. Figure 5–14 illustrates the cost-benefit relationship.

In 1995 the consent order participants began a series of technical meetings to resolve outstanding differences. The foundation of these meetings was the Total Maximum Daily Limit (TMDL) approach, which provided the basis for a phased,

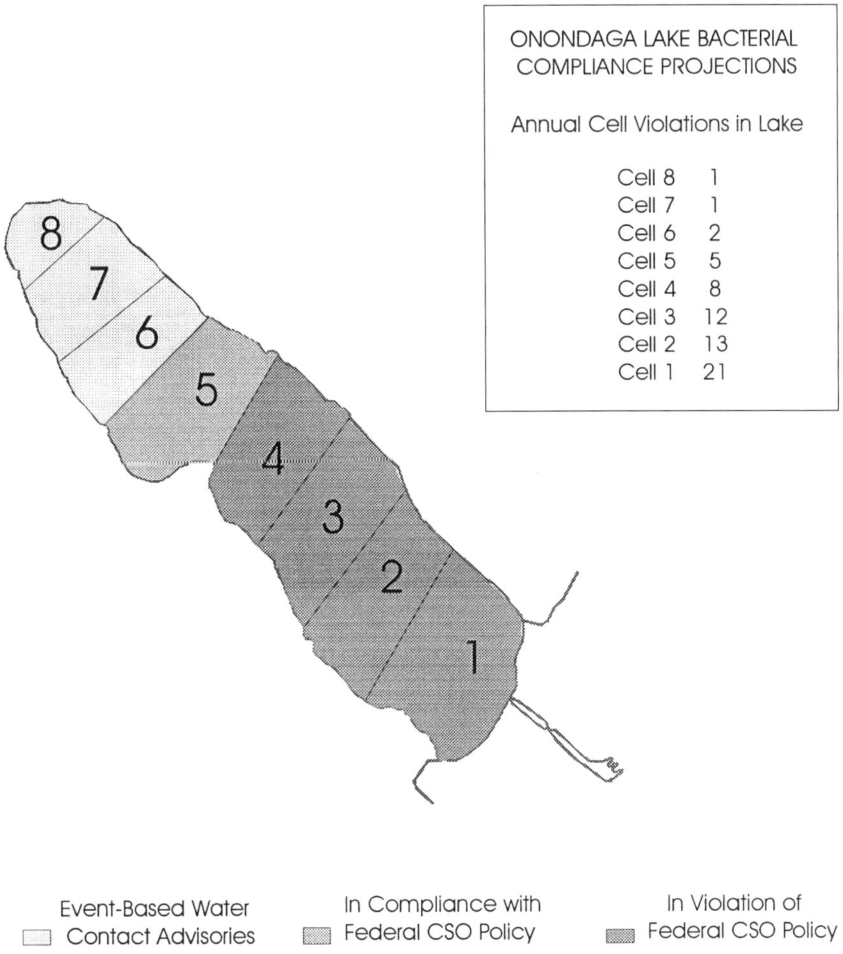

Figure 5–13. Bacteria violations in Onondaga Lake in present condition, Onondaga County combined sewer overflow abatement plan.

"build-and-measure" program. High cost/benefit projects and projects designed to demonstrate innovative technologies were programmed for the initial phase.

Initial meetings served only to emphasize differences, particularly on the matter of storage, until the county presented tables to illustrate the storage feature of the vortex/high-rate disinfection facilities. Table 5–4 illustrates the components of the system, both existing and proposed, and the total wet-weather capture. The technical meeting at which this was presented was decisive in that the proposed regional facilities would ultimately meet the "presumption" clause of the federal CSO policy. The next step in the process was to phase the con-

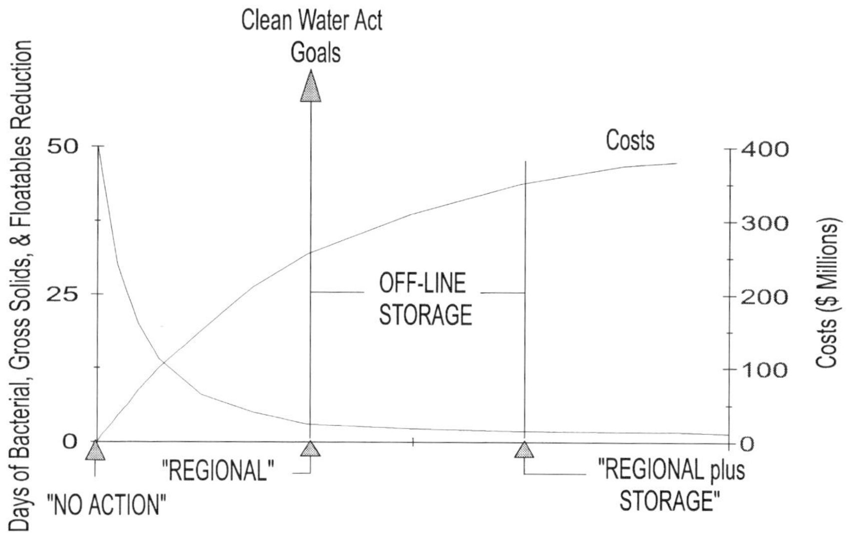

Figure 5-14. Cost-benefit relationship for CSO abatement alternatives, Onondaga County, New York.

Table 5-4. Wet-Weather Combined Sewer System Flows, Syracuse, New York

Regional Facility	Total Wet-Weather Flow in CSS (MG)	Average Annual Interceptor Flow (MG)	Regional Facility CSO Discharge (MG)	RTF Captured Flow (MG)	Percent Capture
		Interim Projects			
Harbor Brook FBM	458.39	331.98	37.92	88.49	92%
Hiawatha RTF	57.90	33.50	0.62	23.78	99%
Teall FCF	84.19	76.79	7.40	0.00	91%
EBSS	284.32	0.00	55.96	228.36	80%
		Intermediate Projects			
Newell RTF	7.22	3.62	1.70	1.90	76%
Midland RTF	574.59	275.39	112.40	186.80	80%
Clinton RTF	543.58	401.68	86.48	55.42	84%
Franklin FCF	527.11	443.81	77.29	6.01	85%
Maltbie FCF	67.79	40.89	21.54	5.36	68%
TOTAL	2605.09	1607.66	401.31	596.12	85%

struction of the facilities over a period of time to provide both a reasonable compliance schedule and an affordable cash flow for the Syracuse Metropolitan Sanitary District. Compliance was necessary in terms of floatables and bacteria. Figure 5–15 illustrates the schedule of projects totaling $144 million (ENR = 5830, 1997) over a period of 15 years and the degree to which compliance will be achieved over that period.

Innovative Projects

Inasmuch as the schedule for the program will be over a period of 15 years, provision was made to incorporate certain innovative projects into the schedule on the basis of potential cost savings and reduction of environmental impact. The three major projects are identified as follows:

Project	**Potential Benefit**
▪ Hiawatha RFT Demonstration	▪ Cost Savings
▪ Newell St. Disinfection Demonstration	▪ Environmental
▪ Harbor Brook Equiflow System	▪ Cost Savings & Environmental

RTF = Regional Treatment Facility

The Hiawatha RTF is the only facility proposed that includes a storage tank in addition to the vortex facility and disinfection tank. All other sites include land for storage should additional pollutant removal, not evident or justified at this time, be required in the future. The Hiawatha Facility will demonstrate whether storage offers any measurable benefit and will illustrate the cost/benefit relationship.

In recognition of the concerns for chlorine toxicity, the Newell Street Facility will demonstrate full-scale alternatives to chlorine/dechlorination. Onsite generation of chlorine dioxide and ultraviolet disinfection will be demonstrated on CSO flows pretreated by an existing vortex facility. The outcome of this project will determine the disinfection technology to be employed at the remaining facilities.

The Harbor Brook Flow Balance Method (FBM) Project is perhaps the most unique innovative project and represents a potential savings of over $50 million. It will detain the entire watershed flow during a CSO event for subsequent treatment at the STP. Figures 5–16 and 5–17 illustrate the watershed components and operations of the FBM respectively.

Effectiveness Evaluation

Evaluation of the technologies as they are put into operation and changes in receiving-water quality will be monitored throughout the 15-year period. Figure 5–18 illustrates the location and components of the receiving-water program.

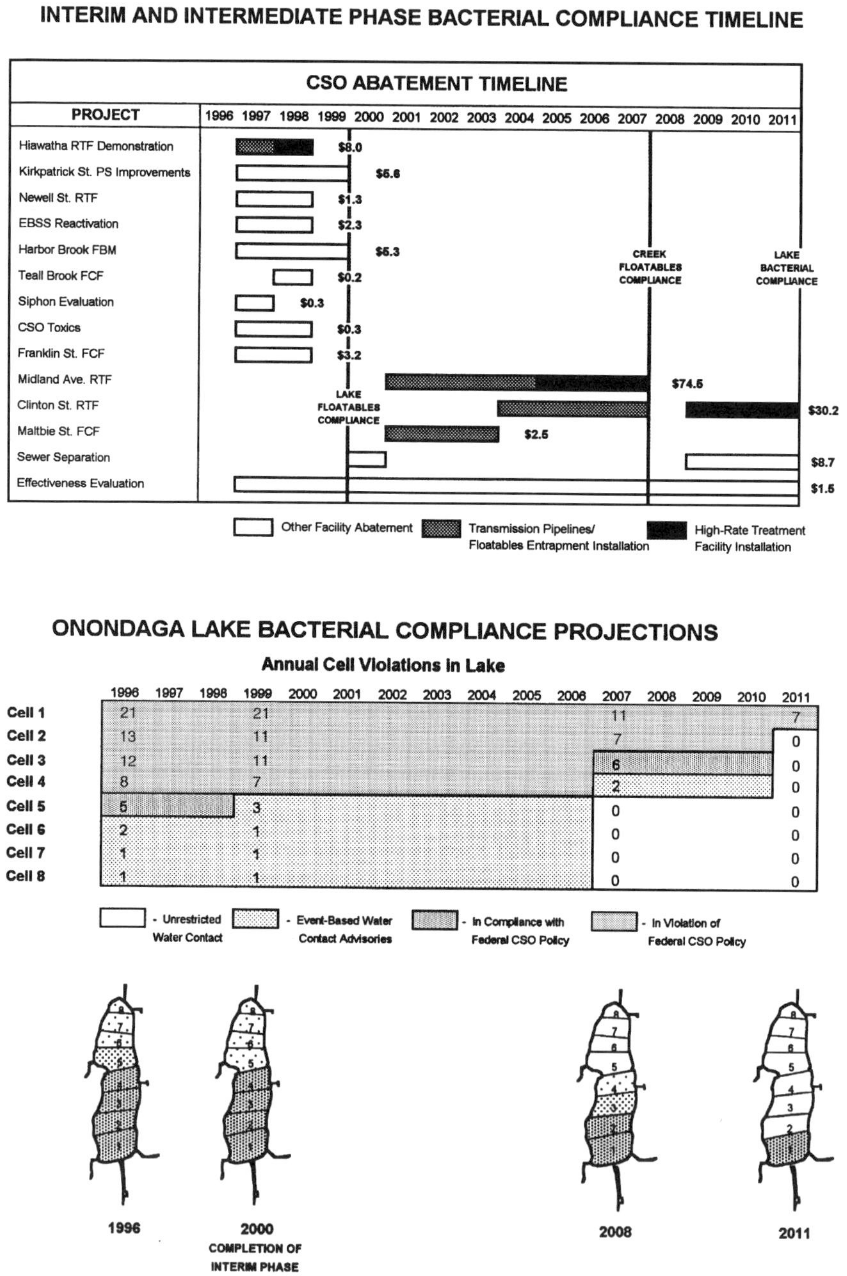

Figure 5-15. CSO abatement timeline and Onondaga bacterial compliance projections.

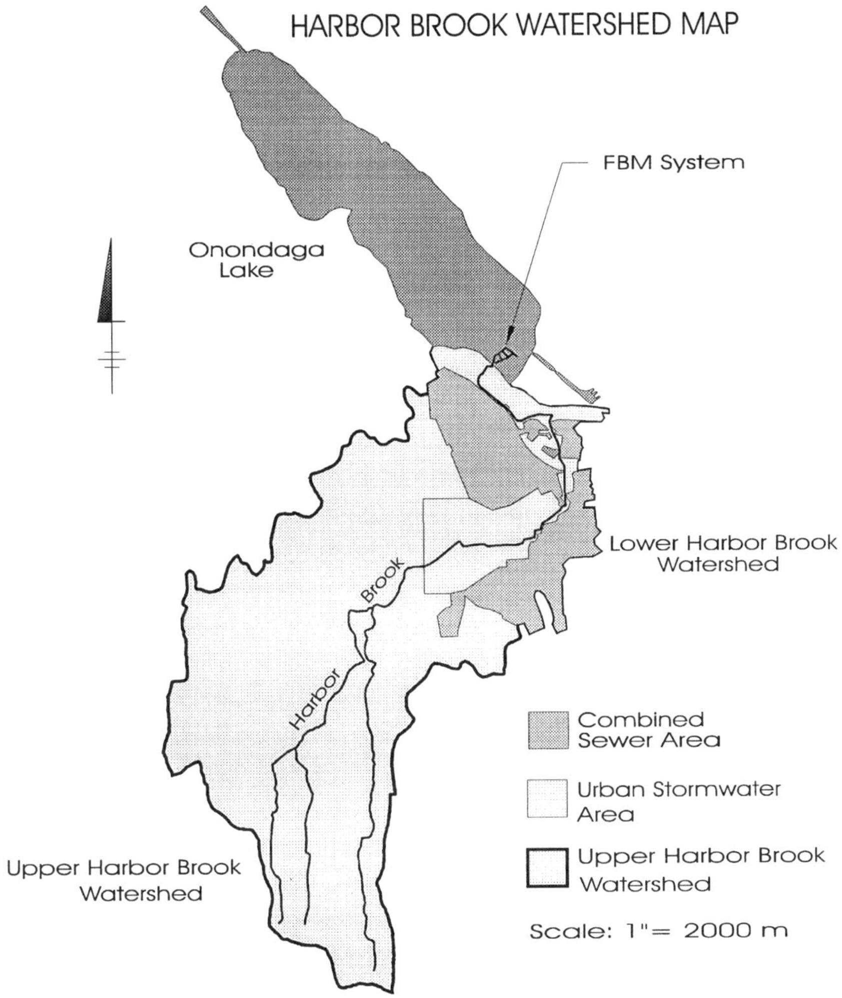

Figure 5-16. Harbor Brook watershed map.

The CSO abatement program represents the culmination of input from all segments of the community and reflects both the community's concern for affordability and its desire to recover a valuable water resource.

Washington, D.C.

Approximately 35% of the city's 12,500 acres is served by combined sewers. Water-quality studies of the three receiving streams indicated that the most

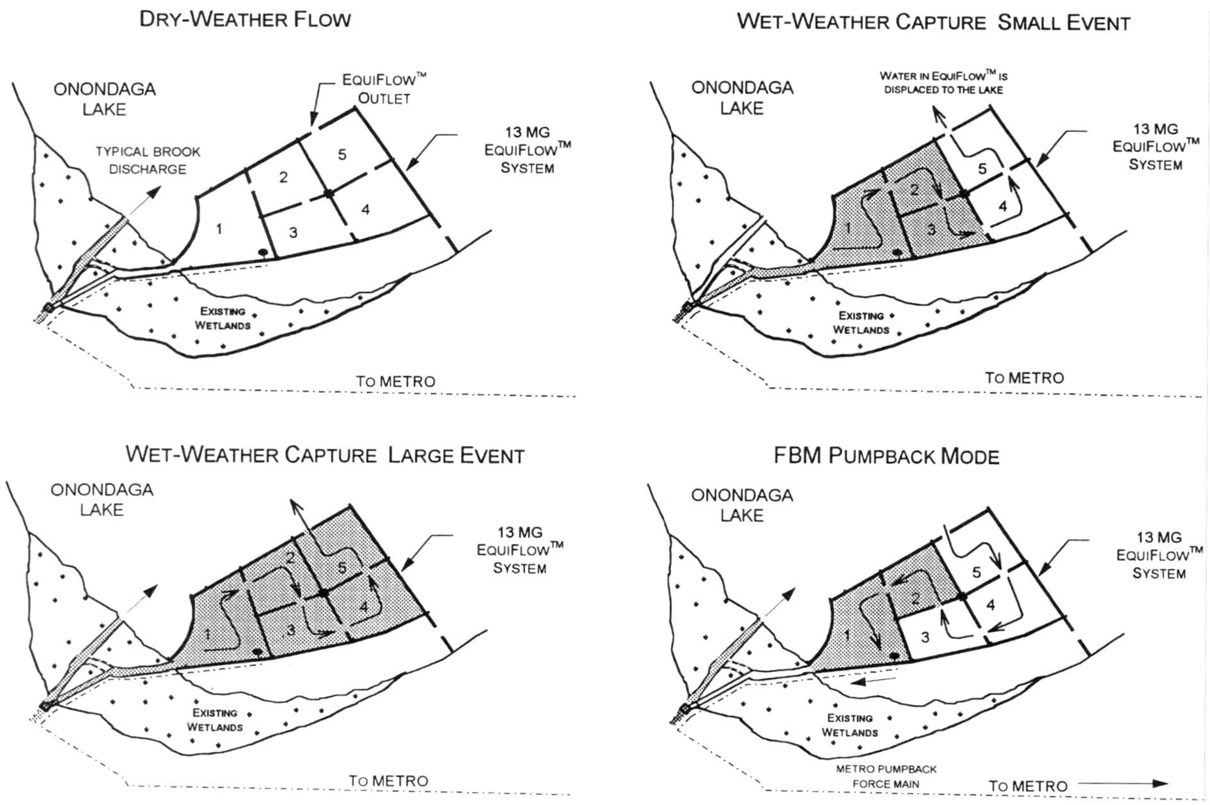

Figure 5-17. Harbor Brook FBM system operational modes.

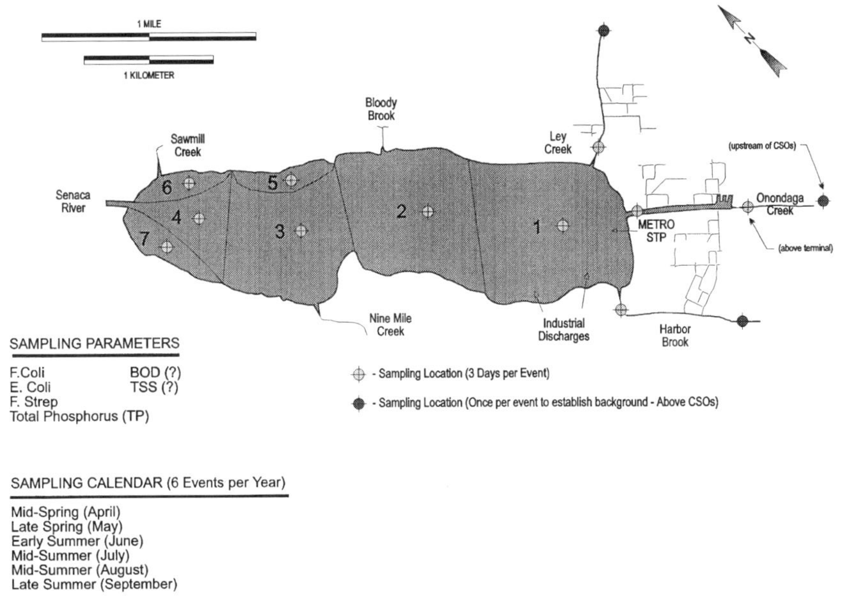

Figure 5–18. Location and components of the receiving-water sampling program.

severely impacted is the Anacostia River; this is because of disproportionate CSO volume and stagnant flow conditions, as outlined in the Facility Planning Study (O'Brien and Gere 1983). Figures 5–19 and 5–20 portray the relative costs of various treatment options for improvements in dissolved oxygen and bacteria concentrations, respectively. Based on the evaluations, a two-segment abatement program was developed that consists of a mix of both collection-system optimization and capital-intensive measures. The first segment of the plan consists of (1) minimal modification of the existing sewage system to maximize conveyance system capacity, (2) a telemetry system to control overflows, and (3) the construction of a 400-mgd end-of-pipe treatment facility near the Robert Kennedy Memorial Stadium for remediation of particular water-quality problems in the Anacostia River. This facility consists of three 57-ft diameter swirl concentrators, followed by high-rate disinfection and postkill neutralization. The final step of the first segment will be a reassessment of environmental conditions following completion of this segment.

In 1991, construction of the three 57-ft-diameter units was completed, resulting in an overall removal of 37% TSS. This percent removal was less than expected from preliminary sampling of the CSO system. It was found that the average TSS at the site of 231 mg/l was considerably less than the TSS found

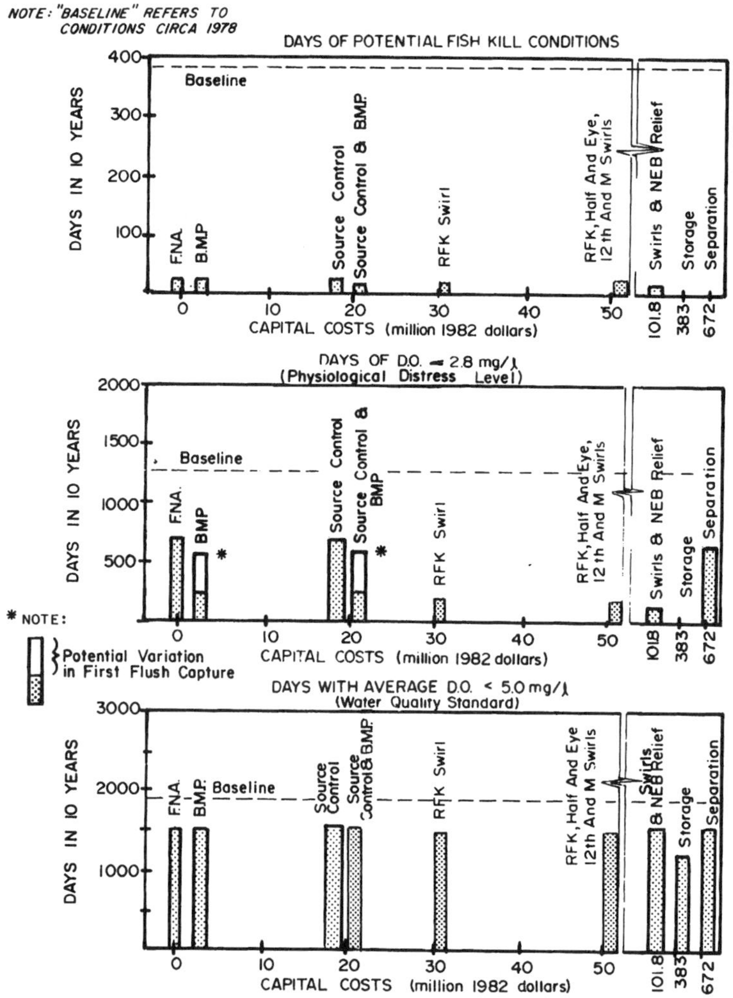

Figure 5–19. Dissolved oxygen impact versus cost for Washington, D.C. (O'Brien & Gere 1983).

during the preliminary sampling. Therefore, the swirl concentrators removed fewer of the solids. As a general rule, the lower the TSS influent to a vortex device, the lower the percent removal; the higher the TSS influent, the higher the percent removal. The Washington, D.C., swirls effectively remove the floatables and remove the percent TSS as predicted from the EPA design curves.

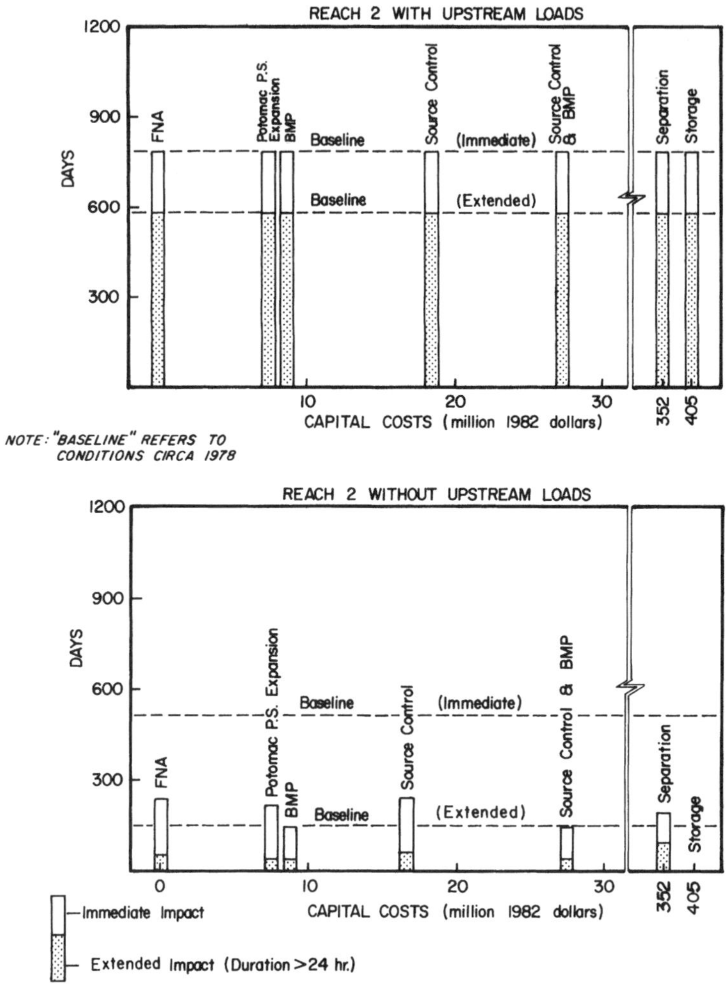

Figure 5–20. Bacterial reductions versus cost for Washington, D.C. (O'Brien & Gere 1983).

References: Chapter 5

Association of Metropolitan Sewerage Agencies (Draft USEPA CSO Permitting Strategy, August 1988), Bulletin No. GB88–22.

Calocerinos & Spina. 1983. "In-water combined sewer overflow abatement study." Report to Onondaga County, NY.

———. 1988. "Hartford combined sewer overflow facility plan." Report to the Metropolitan District Commission, Hartford, CT.

Moffa & Associates. 1988 (ongoing). "Combined sewer overflow abatement facilities plan update." Report to Onondaga County, NY.

Moffa, P. E., J. C. Byron, S. D. Freedman, J. M. Karanik, and R. Ott. 1980. "Methodology for evaluating the impact and abatement of combined sewer overflows: A case study of Onondaga Lake, NY," EPA-600/880-048.

O'Brien & Gere. 1979. "Combined sewer overflow abatement program." Report to Onondaga County, NY.

———. 1983. "Combined sewer overflow abatement program, District of Columbia."

Stearns and Wheler. 1979. "Onondaga Lake storms impact study." Report to Onondaga County, NY.

Appendix **A**

Environmental Protection Agency

40 CFR Part 122

Combined Sewer Overflow (CSO) Control Policy

AGENCY: Environmental Protection Agency (EPA)

ACTION: Final Policy

SUMMARY: EPA has issued a national policy statement entitled "Combined Sewer Overflow (CSO) Control Policy." This policy establishes a consistent national approach for controlling discharges from CSOs to the Nation's waters through the National Pollutant Discharge Elimination System (NPDES) permit program.

FOR FURTHER INFORMATION CONTACT: Jeffrey Lape, Office of Wastewater Enforcement and Compliance, MC-4201, U.S. Environmental Protection Agency, 401 M Street S.W., Washington, DC 20460, (202) 260-7361.

SUPPLEMENTARY INFORMATION: The main purposes of the CSO Control Policy are to elaborate on the Environmental Protection Agency's (EPA's) National (CSO) Control Strategy published on September 8, 1989, at 54 FR 37370, and to expedite compliance with the requirements of the Clean Water Act (CWA). While implementation of the 1989 Strategy has resulted in progress toward controlling CSOs, significant public health and water quality risks remain.

This Policy provides guidance to permittees with CSOs, NPDES authorities, and State water quality standards authorities on coordinating the planning, selection, and implementation of CSO controls that meet the requirements of the CWA and allow for public involvement during the decision-making process. Contained in the Policy are provisions for developing appropriate, site-specific NPDES permit requirements for all combined sewer systems (CSS) that overflow as a result of wet weather events. For example, the Policy lays out two alternative approaches—the "demonstration" and the "presumption" approaches—that provide communities with targets for CSO controls that achieve compliance with the Act, particularly protection of water quality and designated uses. The Policy also includes enforcement initiatives to require the immediate elimination of overflows that occur during dry weather and to ensure that the remaining CWA requirements are complied with as soon as practicable.

The permitting provisions of the Policy were developed as a result of extensive input received from key stakeholders during a negotiated policy dialogue. The CSO stakeholders included representatives from States, environmental groups, municipal organizations, and others. The negotiated dialogue was conducted during the Summer of 1992 by the Office of Water and the Office of Water's Management Advisory Group. The enforcement initiatives, including one which is under way to address CSOs during dry weather, were developed by EPA's Office of Water and Office of Enforcement.

EPA issued a Notice of Availability on the draft CSO Control Policy on January 19, 1993 (58 FR 4994), and requested comments on the draft Policy by March 22, 1993. Approximately 41 sets of written comments were submitted by a variety of interest groups including cities and municipal groups, environmental groups, States, professional organizations, and others. All comments were considered as EPA prepared the Final Policy. The public comments were largely supportive of the draft Policy. EPA received broad endorsement of and support for the key principles and provisions from most commenters. Thus, this final Policy does not include significant changes to the major provisions of the draft Policy, but rather it includes clarification and better explanation of the elements of the Policy to address several of the questions that were raised in the comments. Persons wishing to obtain copies of the public comments or EPA's summary analysis of the comments may write or call the EPA contact person.

The CSO Policy represents a comprehensive national strategy to ensure that municipalities, permitting authorities, water quality standards authorities, and the public engage in a comprehensive and coordinated planning effort to achieve cost-effective CSO controls that ultimately meet appropriate health and environmental objectives. The Policy recognizes the site-specific nature of CSOs and their impacts and provides the necessary flexibility to tailor controls to local situations. Major elements of the Policy ensure that CSO controls are cost-effective and meet the objectives and requirements of the CWA.

The major provisions of the Policy are as follows:

CSO permittees should immediately undertake a process to accurately characterize their CSS and CSO discharges, demonstrate implementation of minimum technology-based controls identified in the Policy, and develop long-term CSO control plans which evaluate alternatives for attaining compliance with the

CWA, including compliance with water quality standards and protection of designated uses. Once the long-term CSO control plans are completed, permittees will be responsible to implement the plans' recommendations as soon as practicable.

State water quality standards authorities will be involved in the long-term CSO control planning effort as well. The water quality standards authorities will help ensure that development of the CSO permittees' long-term CSO control plans are coordinated with the review and possible revision of water quality standards on CSO-impacted waters.

NPDES authorities will issue/reissue or modify permits, as appropriate, to require compliance with the technology-based and water quality–based requirements of the CWA. After completion of the long-term CSO control plan, NPDES permits will be reissued or modified to incorporate the additional requirements specified in the Policy, such as performance standards for the selected controls based on average design conditions, a post-construction water quality assessment program, monitoring for compliance with water quality standards, and a reopener clause authorizing the NPDES authority to reopen and modify the permit if it is determined that the CSO controls fail to meet water quality standards or protect designated uses.

NPDES authorities should commence enforcement actions against permittees that have CWA violations due to CSO discharges during dry weather. In addition, NPDES authorities should ensure the implementation of the minimum technology-based controls and incorporate a schedule into an appropriate enforceable mechanism, with appropriate milestone dates, to implement the required long-term CSO control plan. Schedules for implementation of the long-term CSO control plan may be phased based on the relative importance of adverse impacts upon water quality standards and designated uses, and on a permittee's financial capability.

EPA is developing extensive guidance to support the Policy and will announce the availability of the guidances and other outreach efforts through various means as they become available. For example, EPA is preparing guidance on the nine minimum controls, characterization and monitoring of CSOs, development of long-term CSO control plans, and financial capability.

Permittees will be expected to comply with any existing CSO-related requirements in NPDES permits, consent decrees, or court orders unless revised to be consistent with this Policy.

The policy is organized as follows:

I. INTRODUCTION
   A. Purpose and Principles
   B. Application of Policy
   C. Effect on Current CSO Control Efforts
   D. Small-System Considerations

List of Subjects in 40 CFR Part 122: Water Pollution Control Authority: Clean Water Act, 33 U.S.C. 1251 *et seq.*

Carol M. Browner
Administrator

COMBINED SEWER OVERFLOW (CSO) CONTROL POLICY

I. INTRODUCTION

A. Purpose and Principles

The main purposes of this Policy are to elaborate on EPA's National Combined Sewer Overflow (CSO) Control Strategy published on September 8, 1989, at 54 FR 37370 (1989 Strategy) and to expedite compliance with the requirements of the Clean Water Act (CWA). While implementation of the 1989 Strategy has resulted in progress toward controlling CSOs, significant water quality risks remain.

A combined sewer system (CSS) is a wastewater collection system owned by a State or municipality (as defined by Section 502[4] of the CWA) which conveys sanitary wastewaters (domestic, commercial, and industrial wastewaters) and storm water through a single-pipe system to a Publicly Owned Treatment Works (POTW) Treatment Plant (as defined in 40 CFR 403.3[p]). A CSO is the discharge from a CSS at a point prior to the POTW Treatment Plant. CSOs are point sources subject to NPDES permit requirements including both technology-based and water quality–based requirements of the CWA. CSOs are not subject to secondary treatment requirements applicable to POTWs.

CSOs consist of mixtures of domestic sewage, industrial and commercial wastewaters, and storm water runoff. CSOs often contain high levels of suspended solids, pathogenic microorganisms, toxic pollutants, floatables, nutrients, oxygen-demanding organic compounds, oil and grease, and other pollutants. CSOs can cause exceedances of water quality standards (WQS). Such exceedances may pose risks to human health, threaten aquatic life and its habitat, and impair the use and enjoyment of the Nation's waterways.

This Policy is intended to provide guidance to permittees with CSOs, National Pollutant Discharge Elimination System (NPDES) permitting authorities, State water quality standards authorities, and enforcement authorities. The purpose of the Policy is to coordinate the planning, selection, design, and implementation of CSO management practices and controls to meet the requirements of the CWA and to involve the public fully during the decision-making process.

This Policy reiterates the objectives of the 1989 Strategy:

1. To ensure that if CSOs occur, they are only as a result of wet weather
2. To bring all wet weather CSO discharge points into compliance with the technology-based and water quality–based requirements of the CWA
3. To minimize water quality, aquatic biota, and human health impacts from CSOs

This CSO Control Policy represents a comprehensive national strategy to ensure that municipalities, permitting authorities, water quality standards authorities, and the public engage in a comprehensive and coordinated planning effort to achieve cost-effective CSO controls that ultimately meet appropriate health and environmental objectives and requirements. The Policy recognizes the site-specific nature of CSOs and their impacts and provides the necessary flexibility to tailor controls to local situations. Four key principles of the Policy ensure that CSO controls are cost-effective and meet the objectives of the CWA. The key principles are

1. providing clear levels of control that would be presumed to meet appropriate health and environmental objectives;
2. providing sufficient flexibility to municipalities, especially financially disadvantaged communities, to consider the site-specific nature of CSOs and to determine the most cost-effective means of reducing pollutants and meeting CWA objectives and requirements;
3. allowing a phased approach to implementation of CSO controls considering a community's financial capability; and
4. reviewing and revision, as appropriate, of water quality standards and their implementation procedures when developing CSO control plans to reflect the site-specific wet weather impacts of CSOs.

This Policy is being issued in support of EPA's regulations and policy initiatives. This Policy is Agency guidance only and does not establish or affect legal rights or obligations. It does not establish a binding norm and is not finally determinative of the issues addressed. Agency decisions in any particular case will be made by applying the law and regulations on the basis of specific facts when permits are issued. The Administration has recommended that the 1994 amendments to the CWA endorse this final Policy.

B. Application of Policy

The permitting provisions of this Policy apply to all CSSs that overflow as a result of storm water flow, including snow melt runoff (40 CFR Section 122.26[b][13]). Discharges from CSSs during dry weather are prohibited by the CWA. Accordingly, the permitting provisions of this Policy do not apply to CSOs during dry weather. Dry weather flow is the flow in a combined sewer that results from domestic sewage, groundwater infiltration, commercial and industrial wastewaters, and any other non-precipitation-related flows (e.g., tidal infiltration). In addition to the permitting provisions, the Enforcement and Compliance section of this Policy describes an enforcement initiative being developed for overflows that occur during dry weather.

Consistent with the 1989 Strategy, 30 States that submitted CSO permitting strategies have received EPA approval or, in the case of one State, conditional approval of its strategy. States and EPA Regional Offices should review these strategies and negotiate appropriate revisions to them to implement this Policy. Permitting authorities are encouraged to evaluate water pollution control needs on a watershed management basis and coordinate CSO control efforts with other point and nonpoint source control activities.

C. Effect on Current CSO Control Efforts

EPA recognizes that extensive work has been done by many Regions, States, and municipalities to abate CSOs. As such, portions of this Policy may already have been addressed by permittees' previous efforts to control CSOs. Therefore, portions of this Policy may not apply, as determined by the permitting authority on a case-by-case basis, under the following circumstances:

1. Any permittee that, on the date of publication of this final Policy, has completed or substantially completed construction of CSO control facilities that are designed to meet WQS and protect designated uses, and where it has been determined that WQS are being or will be attained, is not covered by the initial planning and construction provisions in this Policy; however, the operational plan and postconstruction monitoring provisions continue to apply. If, after monitoring, it is determined that WQS are not being attained, the permittee should be required to submit a revised CSO control plan that, once implemented, will attain WQS.
2. Any permittee that, on the date of publication of this final Policy, has substantially developed or is implementing a CSO control program pursuant to an existing permit or enforcement order, and such program is considered by the NPDES permitting authority to be adequate to meet WQS and protect designated uses and is reasonably equivalent to the treatment objectives of this Policy, should complete those facilities without further planning activities otherwise expected by this Policy. Such programs, however, should be reviewed and modified to be consistent with the sensitive area, financial capability, and postconstruction monitoring provisions of this Policy.
3. Any permittee that has previously constructed CSO control facilities in an effort to comply with WQS but has failed to meet such applicable standards or to protect designated uses due to remaining CSOs may receive consideration for such efforts in future permits or enforceable orders for long-term CSO control planning, design, and implementation.

In the case of any ongoing or substantially completed CSO control effort, the NPDES permit or other enforceable mechanism, as appropriate, should be revised to include all appropriate permit requirements consistent with Section IV.B of this Policy.

D. Small-System Considerations

The scope of the long-term CSO control plan, including the characterization, monitoring and modeling, and evaluation of alternatives portions of this Policy, may be difficult for some small CSSs. At the discretion of the NPDES Authority, jurisdictions with populations under 75,000 may not need to complete each of the formal steps outlined in Section II.C of this Policy but should be required through their permits or other enforceable mechanisms to comply with the nine minimum controls (II.B), public participation (II.C.2), and sensitive areas (II.C.3) portions of this Policy. In addition, the permittee may propose to implement any of the criteria contained in this Policy for evaluation of alternatives described in II.C.4. Following approval of the proposed plan, such jurisdictions should construct the control projects and propose a monitoring program sufficient to determine whether WQS are attained and designated uses are protected.

In developing a long-term CSO control plan based on the small-system considerations discussed in the preceding paragraph, permittees are encouraged to discuss the scope of their long-term CSO control plan with the WQS authority and the NPDES authority. These discussions will ensure that the plan includes sufficient information to enable the permitting authority to identify the appropriate CSO controls.

E. Implementation Responsibilities

NPDES authorities (authorized States or EPA Regional Offices, as appropriate) are responsible for implementing this Policy. It is their responsibility to assure that CSO permittees develop long-term CSO control plans and that NPDES permits meet the requirements of the CWA. Further, they are responsible for coordinating the review of the long-term CSO control plan and the development of the permit with the WQS authority to determine if revisions to the WQS are appropriate. In addition, they should determine the appropriate vehicle (i.e., permit reissuance, information request under CWA Section 308 or State equivalent, or enforcement action) to ensure that compliance with the CWA is achieved as soon as practicable.

Permittees are responsible for documenting the implementation of the nine minimum controls and developing and implementing a long-term CSO control plan, as described in this Policy. EPA recognizes that financial considerations are a major factor affecting the implementation of CSO controls. For that reason, this Policy allows consideration of a permittee's financial capability in connection with the long-term CSO control planning effort, WQS review, and negotiation of enforceable schedules. However, each permittee is ultimately responsible for aggressively pursuing financial arrangements for the implementation of its long-term CSO control plan. As part of this effort, communities should apply to their State Revolving Fund program, or other assistance programs as appropriate, for financial assistance.

EPA and the States will undertake action to assure that all permittees with CSSs are subject to a consistent review in the permit development process, have permit requirements that achieve compliance with the CWA, and are subject to enforceable schedules that require the earliest practicable compliance dates considering physical and financial feasibility.

F. Policy Development

This Policy devotes a separate section to each step involved in developing and implementing CSO controls. This is not to imply that each function occurs separately. Rather, the entire process surrounding CSO controls, community planning, WQS and permit development/revision, enforcement/compliance actions, and public participation must be coordinated to control CSOs effectively. Permittees and permitting authorities are encouraged to consider innovative and alternative approaches and technologies that achieve the objectives of this Policy and the CWA.

In developing this Policy, EPA has included information on what responsible parties are expected to accomplish. Subsequent documents will provide additional guidance on how the objectives of this Policy should be met. These documents will provide further guidance on the following: CSO permit writing, the nine minimum controls, long-term CSO control plans, financial capability, sewer system characterization and receiving-water monitoring and modeling, and application of WQS to CSO-impacted waters. For most CSO control efforts, however, sufficient detail has been included in this Policy to begin immediate implementation of its provisions.

II. EPA OBJECTIVES FOR PERMITTEES

A. Overview

Permittees with CSSs that have CSOs should immediately undertake a process to accurately characterize their sewer systems, to demonstrate implementation of the nine minimum controls, and to develop a long-term CSO control plan.

B. Implementation of the Nine Minimum Controls

Permittees with CSOs should submit appropriate documentation demonstrating implementation of the nine minimum controls, including any proposed schedules for completing minor construction activities. The nine minimum controls are

1. proper operation and regular maintenance programs for the sewer system and the CSOs;
2. maximum use of the collection system for storage;
3. review and modification of pretreatment requirements to assure CSO impacts are minimized;
4. maximization of flow to the POTW for treatment;
5. prohibition of CSOs during dry weather;
6. control of solid and floatable materials in CSOs;
7. pollution prevention;
8. public notification to ensure that the public receives adequate notification of CSO occurrences and CSO impacts; and
9. monitoring to effectively characterize CSO impacts and the efficacy of CSO controls.

Selection and implementation of actual control measures should be based on site-specific considerations including the specific CSR's characteristics discussed under the sewer system characterization and monitoring portions of this Policy. Documentation of the nine minimum controls may include operation and maintenance plans, revised sewer use ordinances for industrial users, sewer system inspection reports, infiltration/inflow studies, pollution prevention programs, public notification plans, and facility plans for maximizing the capacities of the existing collection, storage, and treatment systems, as well as contracts and schedules for minor construction programs for improving the existing system's operation. The permittee should also submit any information or data on the degree to which the nine minimum controls achieve compliance with water quality standards. These data and information should include results made available through monitoring and modeling activities done in conjunction with the development of the long-term CSO control plan described in this Policy.

This documentation should be submitted as soon as practicable but no later than two years after the requirement to submit such documentation is included in an NPDES permit or other enforceable mechanism. Implementation of the nine minimum controls with appropriate documentation should be completed as soon as practicable but no later than January 1, 1997. These dates should be included in an appropriate enforceable mechanism.

Because the CWA requires immediate compliance with technology-based controls (Section 301[b]), which on a Best Professional Judgment basis should include the nine minimum controls, a compliance schedule for implementing the nine minimum controls, if necessary, should be included in an appropriate enforceable mechanism.

C. Long-Term CSO Control Plan

Permittees with CSOs are responsible for developing and implementing long-term CSO control plans that will ultimately result in compliance with the requirements of the CWA. The long-term plans should consider the site-specific nature of CSOs and evaluate the cost-effectiveness of a range of control options/strategies. The development of the long-term CSO control plan and its subsequent implementation should also be coordinated with the NPDES authority and the State authority responsible for reviewing and revising the State's WQS. The selected controls should be designed to allow cost-effective expansion or cost-effective retrofitting if additional controls are subsequently determined to be necessary to meet WQS, including existing and designated uses.

This policy identifies EPA's major objectives for the long-term CSO control plan. Permittees should develop and submit this long-term CSO control plan as soon as practicable but generally within two years after the date of the NPDES permit provision, Section 308 information request, or enforcement action requiring the permittee to develop the plan. NPDES authorities may establish a longer timetable for completion of the long-term CSO control plan on a case-by-case basis to account for site-specific factors which may influence the complexity of the planning process. Once agreed upon, these dates should be included in an appropriate enforceable mechanism.

EPA expects each long-term CSO control plan to utilize appropriate information to address the following minimum elements. The Plan should also include both fixed-date project implementation schedules (which may be phased) and a financing plan to design and construct the project as soon as practicable. The minimum elements of the long-term CSO control plan are described below.

1. Characterization, Monitoring, and Modeling of the Combined Sewer System In order to design a CSO control plan adequate to meet the requirements of the CWA, a permittee should have a thorough understanding of its sewer system, the response of the system to various precipitation events, the characteristics of the overflows, and the water quality impacts that result from CSOs. The permittee should adequately characterize through monitoring, modeling, and other means as appropriate, for a range of storm events, the response of its sewer system to wet weather events including the number, location, and frequency of CSOs; volume, concentration, and mass of pollutants discharged; and the impacts of the CSOs on the receiving waters and their designated uses. The permittee may need to consider information on the contribution and importance of other pollution sources in order to develop a final plan designed to meet water quality standards. The purpose of the system characterization, monitoring, and modeling program initially is to assist the permittee in developing appropriate measures to implement the nine minimum controls and, if necessary, to support development of the long-term CSO control plan. The monitoring and modeling data also will be used to evaluate the expected effectiveness of both the nine minimum controls and, if necessary, the long-term CSO controls to meet WQS.

The major elements of a sewer system characterization are described below:

a. Rainfall Records—The permittee should examine the complete rainfall record for the geographic area of its existing CSS using sound statistical procedures and best available data. The permittee should evaluate flow variations in the receiving water body to correlate between CSOs and receiving-water conditions.
b. Combined Sewer System Characterization—The permittee should evaluate the nature and extent of its sewer system through evaluation of available sewer system records, field inspections, and other activities necessary to understand the number, location, and frequency of overflows and their location relative to sensitive areas and to pollution sources in the collection system, such as indirect significant industrial uscrs.
c. CSO Monitoring—The permittee should develop a comprehensive, representative monitoring program that measures the frequency, duration, flow rate, volume, and pollutant concentration of CSO discharges and assesses the impact of the CSOs on the receiving waters. The monitoring program should include necessary CSO effluent and ambient in-stream monitoring and, where appropriate, other monitoring protocols such as biological assessment, toxicity testing, and sediment sampling. Monitoring parameters should include, for example, oxygen-demanding pollutants, nutrients, toxic pollutants, sediment contaminants, pathogens, bacteriological indicators (e.g., Enterococcus, E. Coli), and toxicity. A representative sample of overflow points can be selected that is sufficient to allow characterization of CSO discharges and their water quality impacts and to facilitate evaluation of control plan alternatives.
d. Modeling—Modeling of a sewer system is recognized as a valuable tool for predicting sewer system response to various wet weather events and assessing water quality impacts when evaluating different control strategies and alternatives. EPA supports the proper and effective use of models, where appropriate, in the evaluation of the nine minimum controls and the development of the long-term CSO control plan. It is also recognized that there are many models which may be used to do this. These models range from simple to complex. Having decided to USQ a model, the permittee should base its choice of a model on the characteristics of its sewer system, the number and location of overflow points, and the sensitivity of the receiving water body to the CSO discharges. Use of models should include appropriate calibration and verification with field measurements. The sophistication of the model should relate to the complexity of the system to be modeled and to the information needs associated with evaluation of CSO control options and water quality impacts. EPA believes that continuous simulation models, using historical rainfall data, may be the best way to model sewer systems, CSOs, and their impacts. Because of the iterative nature of modeling

sewer systems, CSOs, and their impacts, monitoring and modeling efforts are complementary and should be coordinated.

2. Public Participation In developing its long-term CSO control plan, the permittee will employ a public-participation process that actively involves the affected public in the decision-making to select the long-term CSO controls. The affected public includes rate payers, industrial users of the sewer system, persons who reside downstream from the CSOs, persons who use and enjoy these downstream waters, and any other interested persons.

3. Consideration of Sensitive Areas EPA expects a permittee's long-term CSO control plan to give the highest priority to controlling overflows to sensitive areas. Sensitive areas, as determined by the NPDES authority in coordination with State and Federal agencies, as appropriate, include designated Outstanding National Resource Waters, National Marine Sanctuaries, waters with threatened or endangered species and their habitat, waters with primary-contact recreation, public drinking water intakes or their designated protection areas, and shellfish beds. For such areas, the long-term CSO control plan should

a. prohibit new or significantly increased overflows;
b. i. eliminate or relocate overflows that discharge to sensitive areas wherever physically possible and economically achievable, except where elimination or relocation would provide less environmental protection than additional treatment; or
   ii. where elimination or relocation is not physically possible and economically achievable, or would provide less environmental protection than additional treatment, provide the level of treatment for remaining overflows deemed necessary to meet WQS for full protection of existing and designated uses (in any event, the level of control should not be less than those described in Evaluation of Alternatives below); and
c. where elimination or relocation has been proven not to be physically possible and economically achievable, require, for each subsequent permit term, a reassessment based on new or improved techniques to eliminate or relocate, or on changed circumstances that influence economic achievability.

4. Evaluation of Alternatives EPA expects the long-term CSO control plan to consider a reasonable range of alternatives. The plan should, for example, evaluate controls that would be necessary to achieve zero overflow events per year or an average of one to three, four to seven, or eight to twelve overflow events per year. Alternatively, the long-term plan could evaluate controls that achieve 100% capture, 90% capture, 85% capture, 80% capture, and 75% capture for treatment.

The long-term control plan should also consider expansion of POTW secondary and primary capacity in the CSO abatement alternative analysis. The analysis of alternatives should be sufficient to make a reasonable assessment of cost and performance as described in Section II.C.5. Because the final long-term CSO control plan will become the basis for NPDES permit limits and requirements, the selected controls should be sufficient to meet CWA requirements.

In addition to considering sensitive areas, the long-term CSO control plan should adopt one of the following approaches:

a. "Presumption" Approach A program that meets any of the criteria listed below would be presumed to provide an adequate level of control to meet the water quality–based requirements of the CWA, provided the permitting authority determines that such presumption is reasonable in light of the data and analysis conducted in the characterization, monitoring, and modeling of the system and the consideration of sensitive areas described above. The following criteria are provided because data and modeling of wet weather events often do not give a clear picture of the level of CSO controls necessary to protect WQS:

i. no more than an average of four overflow events per year, provided that the permitting authority may allow up to two additional overflow events per year (for the purpose of this criterion, an overflow event is one or more overflows from a CSS as the result of a precipitation event that does not receive the minimum treatment specified below); or
ii. the elimination or the capture for treatment of no less than 85% by volume of the combined sewage collected in the CSS during precipitation events on a system-wide annual average basis; or
iii. the elimination or removal of no less than the mass of the pollutants, identified as causing water quality impairment through the sewer system characterization, monitoring, and modeling effort, for the volumes that would be eliminated or captured for treatment under paragraph ii above.

Combined sewer flows remaining after implementation of the nine minimum controls and within the criteria specified at II.C.4.a.i or ii should receive a minimum of

- primary clarification (removal of floatables and settleable solids may be achieved by any combination of treatment technologies or methods that are shown to be equivalent to primary clarification);
- solids and floatables disposal; and
- disinfection of effluent, if necessary, to meet WQS, protect designated uses, and protect human health, including removal of harmful disinfection chemical residuals, where necessary.

b. "Demonstration" Approach A permittee may demonstrate that a selected control program, though not meeting the criteria specified in II.C.4.a above, is adequate to meet the water quality–based requirements of the CWA. The permittee should successfully demonstrate each of the following:

i. the planned control program is adequate to meet WQS and protect designated uses, unless WQS or uses cannot be met as a result of natural background conditions or pollution sources other than CSOs;
ii. the CSO discharges remaining after implementation of the planned control program will not preclude the attainment of WQS or the receiving waters' designated uses or contribute to their impairment. Where WQS and designated uses are not met in part because of natural background conditions or pollution sources other than CSOs, a total maximum daily load, including a wasteload allocation and a load allocation, or other means should be used to apportion pollutant loads;
iii. the planned control program will provide the maximum pollution reduction benefits reasonably attainable; and
iv. the planned control program is designed to allow cost-effective expansion or cost-effective retrofitting if additional controls are subsequently determined to be necessary to meet WQS or designated uses.

5. Cost/Performance Considerations The permittee should develop appropriate cost/performance curves to demonstrate the relationships among a comprehensive set of reasonable control alternatives that correspond to the different ranges specified in Section II.C.4. This should include an analysis to determine where the increment of pollution reduction achieved in the receiving water diminishes compared to the increased costs. This analysis, often known as knee of the curve, should be among the considerations used to help guide selection of controls.

6. Operational Plan After agreement between the permittee and NPDES authority on the necessary CSO controls to be implemented under the long-term CSO control plan, the permittee should revise the operation and maintenance program developed as part of the nine minimum controls to include the agreed-upon long-term CSO controls. The revised operation and maintenance program should maximize the removal of pollutants during and after each precipitation event using all available facilities within the collection and treatment system. For any flows in excess of the criteria specified at II.C.4.a.i, ii, or iii and not receiving the treatment specified in II.C.4.a, the operational plan should ensure that such flows receive treatment to the greatest extent practicable.

7. Maximizing Treatment at the Existing POTW Treatment Plant In some communities, POTW treatment plants may have primary treatment capacity in excess of their secondary treatment capacity. One effective strategy to abate pollution resulting from CSOs is to maximize the delivery of flows during wet weather to the POTW treatment plant for treatment. Delivering these flows can have two significant water quality benefits: first, increased flows during wet weather to the POTW treatment plant may enable the permittee to eliminate or minimize overflows to sensitive areas; second, this would maximize the use of available POTW facilities for wet weather flows and would ensure that combined sewer flows receive at least primary treatment prior to discharge.

Under EPA regulations, the intentional diversion of waste streams from any portion of a treatment facility, including secondary treatment, is a bypass. EPA bypass regulations at 40 CFR Section 122.41(m) allow for a facility to bypass some or all the flow from its treatment process under specified limited circumstances. Under the regulation, the permittee must show that the bypass was unavoidable to prevent loss of life, personal injury, or severe property damage, that there was no feasible alternative to the bypass, and that the permittee submitted the required notices. In addition, the regulation provides that a bypass may be approved only after consideration of adverse effects.

Normally, it is the responsibility of the permittee to document, on a case-by-case basis, compliance with 40 CFR Section 122.41(m) in order to bypass flows legally. For some CSO-related permits, the study of feasible alternatives in the control plan may provide sufficient support for the permit record and for approval of a CSO-related bypass in the permit itself, and define the specific parameters under which a bypass can legally occur. For approval of a CSO-related bypass, the long-term CSO control plan, at a minimum, should provide justification for the cutoff point at which the flow will be diverted from the secondary treatment portion of the treatment plant, and provide a benefit-cost analysis demonstrating that conveyance of wet weather flow to the POTW for primary treatment is more beneficial than other CSO abatement alternatives such as storage and pump back for secondary treatment, sewer separation, or satellite treatment. Such a permit must define under what specific wet weather conditions a CSO-related bypass is allowed and also specify what treatment or monitoring and what effluent limitations and requirements apply to the bypass flow. The permit should also provide that approval for the CSO-related bypass will be reviewed and may be modified or terminated if there is a substantial increase in the volume or character of pollutants being introduced to the POTW. The CSO-related bypass provision in the permit should also make it clear that all wet weather flows passing the headworks of the POTW treatment plant will receive at least primary clarification; solids and floatables removal, disposal, and disinfection, where necessary; and any other treatment that can reasonably be provided.

Under this approach, EPA would allow a permit to authorize a CSO-related bypass of the secondary treatment portion of the POTW treatment plant for combined sewer flows in certain identified circumstances. This provision would apply only to those situations where the POTW would ordinarily meet the requirements of 40 CFR 122.41(m) as evaluated on a case-by-case basis. Therefore, there must be sufficient data in the administrative record (reflected in the permit fact sheet or statement of basis) supporting all the requirements in 40 CFR Section 122.41(m)(4) for approval of an anticipated bypass.

For the purposes of applying this regulation to CSO permittees, "severe property damage" could include situations where flows above a certain level wash out the POTW's secondary treatment system. EPA further believes that the feasible alternatives requirement of the regulation can be met if the record shows that the secondary treatment system is properly operated and maintained, that the system has been designed to meet secondary limits for flows greater than the peak dry weather flow, plus an appropriate quantity of wet weather flow, and that it is either technically or financially infeasible to provide secondary treatment at the existing facilities for greater amounts of wet weather flow. The feasible alternative analysis should include, for example, consideration of enhanced primary treatment (e.g., chemical addition) and nonbiological secondary treatment. Other bases supporting a finding of no feasible alternative may also be available on a case-by-case basis. As part of its consideration of possible adverse effects resulting from the bypass, the permitting authority should also ensure that the bypass will not cause exceedances of WQS.

This Policy does not address the appropriateness of approving anticipated bypasses through NPDES permits in advance outside the CSO context.

8. Implementation Schedule The permittee should include all pertinent information in the long-term control plan necessary to develop the construction and financing schedule for implementation of CSO controls. Schedules for implementation of the CSO controls may be phased based on the relative importance of adverse impacts upon WQS and designated uses, priority projects identified in the long-term plan, and a permittee's financial capability.

Construction phasing should consider

a. eliminating overflows that discharge to sensitive areas as the highest priority;
b. use impairment;
c. the permittee's financial capability including consideration of such factors as
   i. median household income;
   ii. total annual wastewater and CSO control costs per household as a percent of median household income;
   iii. overall net debt as a percent of full market property value;
   iv. property tax revenues as a percent of full market property value;
   v. property tax collection rate;
   vi. unemployment; and
   vii. bond rating;

d. grant and loan availability;
e. previous and current residential, commercial, and industrial sewer user fees and rate structures; and
f. other viable funding mechanisms and sources of financing.

9. Postconstruction Compliance Monitoring Program The selected CSO controls should include a postconstruction water quality monitoring program adequate to verify compliance with water quality standards and protection of designated uses as well as to ascertain the effectiveness of CSO controls. This water quality compliance monitoring program should include a plan to be approved by the NPDES authority that details the monitoring protocols to be followed, including the necessary effluent and ambient monitoring and, where appropriate, other monitoring protocols such as biological assessments, whole effluent toxicity testing, and sediment sampling.

III. COORDINATION WITH STATE WATER QUALITY STANDARDS

A. Overview

WQS are State adopted, or Federally promulgated rules which serve as the goals for the water body and the legal basis for the water quality–based NPDES permit requirements under the CWA. WQS consist of uses which States designate for their water bodies, criteria to protect the uses, an antidegradation policy to protect the water quality improvements gained, and other policies affecting the implementation of the standards. A primary objective of the long-term CSO control plan is to meet WQS, including the designated uses, through reducing risks to human health and the environment by eliminating, relocating, or controlling CSOs to the affected waters.

State WQS authorities, NPDES authorities, EPA regional offices, permittees, and the public should meet early and frequently throughout the long-term CSO control planning process. Development of the long-term plan should be coordinated with the review and appropriate revision of WQS and implementation procedures on CSO-impacted waters to ensure that the long-term controls will be sufficient to meet water quality standards. As part of these meetings, participants should agree on the data, information, and analyses needed to support the development of the long-term CSO control plan, the review of applicable WQS, and implementation procedures, if appropriate. Agreements should be reached on the monitoring protocols and models that will be used to evaluate the water quality impacts of the overflows, to analyze the attainability of the WQS, and to determine the water quality–based requirements for the permit. Many opportunities exist for permittees and States to share information as control programs are developed and as WQS are reviewed. Such information should assist States in determining the need for revisions to WQS and implementation procedures to better reflect the site-specific wet weather impacts of CSOs. Coordinating the development of the long-term CSO control plan and the review of the WQS and implementation procedures provides greater assurance that the long-term control plan selected and the limits and requirements included in the NPDES permit will be sufficient to meet WQS and to comply with Sections 301(b)(1)(c) and 402(a)(2) of the CWA.

EPA encourages States and permittees jointly to sponsor workshops for the affected public in the development of the long-term CSO control plan and during the development of appropriate revisions to WQS for CSO-impacted waters. Workshops provide a forum for including the public in discussions of the implications of the proposed long-term CSO control plan on the water quality and uses for the receiving water.

B. Water Quality Standards Reviews

The CWA requires States to periodically, but at least once every three years, hold public hearings for the purpose of reviewing applicable water quality standards and, as appropriate, modifying and adopting standards. States must provide the public an opportunity to comment on any proposed revision to water quality standards, and all revisions must be submitted to EPA for review and approval.

EPA regulations and guidance provide States with the flexibility to adapt their WQS and implementation procedures to reflect site-specific conditions including those related to CSOs. For example, a State may adopt site-specific criteria for a particular pollutant if the State determines that the site-specific criteria fully protects the designated use (40 CFR Section 131.11). In addition, the regulations at 40 CFR Section 131.10(g), (h), and (j) specify when and how a designated use may be modified. A State may remove a designated use from its water quality standards only if the designated use is not an existing use. An existing use is a use actually attained in the water body on or after November 28, 1975. Furthermore, a State may not remove a designated use that will be attained by implementing the technology-based effluent limits required under Sections 301(b) and 306 of the CWA and by implementing cost-effective and reasonable best management practices for nonpoint source controls. Thus, if a State has a reasonable basis to determine that the current designated use could be attained after implementation of the technology-based controls of the CWA, then the use cannot be removed.

In determining whether a use is attainable and prior to removing a designated use, States must conduct and submit to EPA a use attainability analysis. A use attainability analysis is a structured scientific assessment of the factors affecting the use, including the physical, chemical, biological, and economic factors described in 40 CFR Section 131.10(g). As part of the analysis, States should evaluate whether the designated use could be attained if CSO controls were implemented. For example, States should examine if sediment loadings from CSOs could be reduced so as not to bury spawning beds, or if biochemical oxygen-demanding material in the effluent or the toxicity of the effluent could be corrected so as to reduce the acute or chronic physiological stress on or bioaccumulation potential of aquatic organisms.

In reviewing the attainability of their WQS and the applicability of their implementation procedures to CSO-impacted waters, States are encouraged to define more explicitly their recreational and aquatic life uses and then, if appropriate, modify the criteria accordingly to protect the designated uses.

Another option is for States to adopt partial uses by defining when primary-contact recreation such as swimming does not exist, such as during certain seasons of the year in northern climates or during a particular type of storm event. In making such adjustments to their uses, States must ensure that downstream uses are protected and that during other seasons or after the storm event has passed, the use is fully protected.

In addition to defining recreational uses with greater specificity, States are also encouraged to define the aquatic uses more precisely. Rather than "aquatic life use protection," States should consider defining the type of fishery to be protected such as a cold water fishery (e.g., trout or salmon) or a warm weather fishery (e.g., bluegill or largemouth bass). Explicitly defining the type of fishery to be protected may assist the permittee in enlisting the support of citizens for a CSO control plan.

A water quality standard variance may be appropriate, in limited circumstances on CSO-impacted waters, where the State is uncertain as to whether a standard can be attained and time is needed for the State to conduct additional analyses on the attainability of the standard. Variances are short-term modifications in water quality standards. Subject to EPA approval, States, with their own statutory authority, may grant a variance to a specific discharger for a specific pollutant. The justification for a variance is similar to that required for a permanent change in the standard, although the showings needed are less rigorous. Variances are also subject to public-participation requirements of the water quality standards and permits programs and are reviewable generally every three years. A variance allows the CSO permit to be written to meet the "modified" water quality standard as analyses are conducted and as progress is made to improve water quality.

Justifications for variances are the same as those identified in 40 CFR Section 131.10(g) for modifications in uses. States must provide an opportunity for public review and comment on all variances. If States use the permit as the vehicle to grant the variance, notice of the permit must clearly state that the variance modifies the State's water quality standards. If the variance is approved, the State appends the variance to the State's standards and reviews the variance every three years.

IV. EXPECTATIONS FOR PERMITTING AUTHORITIES

A. Overview

CSOs are point sources subject to NPDES permit requirements including both technology-based and water quality–based requirements of the CWA. CSOs are not subject to secondary treatment regulations applicable to publicly owned treatment works (*Montgomery Environmental Coalition vs. Costle,* 646 F.2d 568 [D.C. Cir. 1980]).

All permits for CSOs should require the nine minimum controls as a minimum best-available technology economically achievable and best-conventional technology (BAT/BCT) established on a best-professional judgment (BPJ) basis by the permitting authority (40 CFR Section 125.3). Water quality–based requirements are to be established based on applicable water quality standards.

This policy establishes a uniform, nationally consistent approach to developing and issuing NPDES permits to permittees with CSOs. Permits for CSOs should be developed and issued expeditiously. A single, systemwide permit generally should be issued for all discharges, including CSOs, from a CSS operated by a single authority. When different parts of a single CSS are operated by more than one authority, permits issued to each authority should generally require joint preparation and implementation of the elements of this Policy and should specifically define the responsibilities and duties of each authority. Permittees should be required to coordinate systemwide implementation of the nine minimum controls and the development and implementation of the long-term CSO control plan. The individual authorities are responsible for their own discharges and should cooperate with the permittee for the POTW receiving the flows from the CSS. When a CSO is permitted separately from the POTW, both permits should be cross-referenced for informational purposes.

EPA Regions and States should review the CSO permitting priorities established in the State CSO Permitting Strategies developed in response to the 1989 Strategy. Regions and States may elect to revise these previous priorities. In setting permitting priorities, Regions and States should not just focus on those permittees that have initiated monitoring programs. When setting priorities, Regions and States should consider, for example, the known or potential impact of CSOs on sensitive areas and the extent of upstream industrial user discharges to the CSS.

During the permittee's development of the long-term CSO control plan, the permit writer should promote coordination between the permittee and State WQS authority in connection with possible WQS revisions. Once the permittee has completed development of the long-term CSO control plan and has coordinated with the permitting authority the selection of the controls necessary to meet the requirements of the CWA, the permitting authority should include, in an appropriate enforceable mechanism, requirements for implementation of the long-term

CSO control plan, including conditions for water quality monitoring and operation and maintenance.

B. NPDES Permit Requirements

Following are the major elements of NPDES permits to implement this Policy and ensure protection of water quality:

1. Phase I Permits—Requirements for Demonstration of Nine Minimum Controls and Development of the Long-Term CSO Control Plan In the Phase I permit issued/modified to reflect the Policy, the NPDES authority should at least require permittees to

a. immediately implement BAT/BCT, which at a minimum includes the nine minimum controls, as determined on a BPJ basis by the permitting authority;
b. develop and submit a report documenting the implementation of the nine minimum controls within two years of permit issuance/modification;
c. comply with applicable WQS, no later than the date allowed under the State's WQS, expressed in the form of a narrative limitation; and
d. develop and submit, consistent with this Policy and based on a schedule in an appropriate enforceable mechanism, a long-term CSO control plan as soon as practicable, but generally within two years after the effective date of the permit issuance/modification. However, permitting authorities may establish a longer timetable for completion of the long-term CSO control plan on a case-by-case basis to account for site-specific factors that may influence the complexity of the planning process.

The NPDES authority should include compliance dates on the fastest practicable schedule for each of the nine minimum controls in an appropriate enforceable mechanism issued in conjunction with the Phase I permit. The use of enforceable orders is necessary unless Congress amends the CWA. All orders should require compliance with the nine minimum controls no later than January 1, 1997.

2. Phase II Permits—Requirements for Implementation of a Long-Term CSO Control Plan Once the permittee has completed development of the long-term CSO control plan and the selection of the controls necessary to meet CWA requirements has been coordinated with the permitting and WQS authorities, the permitting authority should include, in an appropriate enforceable mechanism, requirements for implementation of the long-term CSO control plan as soon as practicable. Where the permittee has selected controls based on the "presumption" approach described in Section II.C.4, the permitting authority must have determined that the presumption that such level of treatment will achieve water quality standards is reasonable in light of the data and analysis conducted under this Policy. The Phase II permit should contain

a. requirements to implement the technology-based controls including the nine minimum controls determined on a BPJ basis;
b. narrative requirements which insure that the selected CSO controls are implemented, operated, and maintained as described in the long-term CSO control plan;
c. water quality–based effluent limits under 40 CFR Sections 122.44(d)(1) and 122.44(k), requiring, at a minimum, compliance with, no later than the date allowed under the State's WQS, the numeric performance standards for the selected CSO controls, based on average design conditions specifying at least one of the following:
    i. a maximum number of overflow events per year for specified design conditions consistent with II.C.4.a.i; or
    ii. a minimum percentage capture of combined sewage by volume for treatment under specified design conditions consistent with II.C.4.a.ii; or
    iii. a minimum removal of the mass of pollutants discharged for specified design conditions consistent with II.C.4.a.iii; or
    iv. performance standards and requirements that are consistent with II.C.4.b. of the Policy;
d. a requirement to implement, with an established schedule, the approved postconstruction water quality assessment program including requirements to monitor and collect sufficient information to demonstrate compliance with WQS and protection of designated uses as well as to determine the effectiveness of CSO controls;
e. a requirement to reassess overflows to sensitive areas in those cases where elimination or relocation of the overflows is not physically possible and economically achievable. The reassessment should be based on consideration of new or improved techniques to eliminate or relocate overflows or changed circumstances that influence economic achievability;
f. conditions establishing requirements for maximizing the treatment of wet weather flows at the POTW treatment plant, as appropriate, consistent with Section II.C.7 of this Policy;
g. a reopener clause authorizing the NPDES authority to reopen and modify the permit upon determination that the CSO controls fail to meet WQS or protect designated uses. Upon such determination, the NPDES authority should promptly notify the permittee and proceed to modify or reissue the permit. The permittee should be required to develop, sub-

mit, and implement, as soon as practicable, a revised CSO control plan which contains additional controls to meet WQS and designated uses. If the initial CSO control plan was approved under the demonstration provision of Section II.C.4.b, the revised plan, at a minimum, should provide for controls that satisfy one of the criteria in Section II.C.4.a unless the permittee demonstrates that the revised plan is clearly adequate to meet WQS at a lower cost and it is shown that the additional controls resulting from the criteria in Section II.C.4.a will not result in a greater overall improvement in water quality.

Unless the permittee can comply with all of the requirements of the Phase II permit, the NPDES authority should include, in an enforceable mechanism, compliance dates on the fastest practicable schedule for those activities directly related to meeting the requirements of the CWA. For major permittees, the compliance schedule should be placed in a judicial order. Proper compliance with the schedule for implementing the controls recommended in the long-term CSO control plan constitutes compliance with the elements of this Policy concerning planning and implementation of a long term CSO remedy.

3. Phasing Considerations Implementation of CSO controls may be phased based on the relative importance of and adverse impacts upon WQS and designated uses, as well as the permittee's financial capability and its previous efforts to control CSOs. The NPDES authority should evaluate the proposed implementation schedule and construction phasing discussed in Section II.C.8 of this Policy. The permit should require compliance with the controls proposed in the long-term CSO control plan no later than the applicable deadline(s) under the CWA or State law. If compliance with the Phase II permit is not possible, an enforceable schedule, consistent with the Enforcement and Compliance Section of this Policy, should be issued in conjunction with the Phase II permit which specifies the schedule and milestones for implementation of the long-term CSO control plan.

V. ENFORCEMENT AND COMPLIANCE

A. Overview

It is important that permittees act immediately to take the necessary steps to comply with the CWA. The CSO enforcement effort will commence with an initiative to address CSOs that discharge during dry weather, followed by an enforcement effort in conjunction with permitting CSOs discussed earlier in this Policy. Success of the enforcement effort will depend in large part upon expeditious action by NPDES authorities in issuing enforceable permits that include requirements both for the nine minimum controls and for compliance with all other requirements of the CWA. Priority for enforcement actions should be set based on environmental impacts or sensitive areas affected by CSOs.

As a further inducement for permittees to cooperate with this process, EPA is prepared to exercise its enforcement discretion in determining whether or not to seek civil penalties for past CSO violations if permittees meet the objectives and schedules of this Policy and do not have CSOs during dry weather.

B. Enforcement of CSO Dry Weather Discharge Prohibition

EPA intends to commence immediately an enforcement initiative against CSO permittees which have CWA violations due to CSO during dry weather. Discharges during dry weather have always been prohibited by the NPDES program. Such discharges can create serious public health and water quality problems. EPA will use its CWA Section 308 monitoring, reporting, and inspection authorities, together with NPDES State authorities, to locate these violations and to determine their causes. Appropriate remedies and penalties will be sought for CSOs during dry weather. EPA will provide NPDES authorities more specific guidance on this enforcement initiative separately.

C. Enforcement of Wet Weather CSO Requirements

Under the CWA, EPA can use several enforcement options to address permittees with CSOs. Those options directly applicable to this Policy are Section 308 Information Requests, Section 309(a) Administrative Orders, Section 309(g) Administrative Penalty Orders, Section 309(b) and (d) Civil Judicial Actions, and Section SD4 Emergency Powers. NPDES States should use comparable means.

NPDES authorities should set priorities for enforcement based on environmental impacts or sensitive areas affected by CSOs. Permittees that have voluntarily initiated monitoring and are progressing expeditiously toward appropriate CSO controls should be given due consideration for their efforts.

1. Enforcement for Compliance with Phase I Permits Enforcement for compliance with Phase I permits will focus on requirements to implement at least the nine minimum controls and develop the long-term CSO control plan leading to compliance with the requirements of the CWA. Where immediate compliance with the Phase I permit is infeasible, the NPDES authority should issue an enforceable schedule, in concert with the Phase I permit, requiring compliance with the CWA and imposing compliance schedules with dates for each of the nine minimum controls as soon as practicable. All enforcement authorities should require compliance with the nine minimum controls no later than January 1, 1997. Where the NPDES authority is issuing an order with a compliance schedule for the nine minimum controls, this order should also include a schedule for development of the long-term CSO control plan.

If a CSO permittee fails to meet the final compliance date of the schedule, the NPDES authority should initiate appropriate judicial action.

2. Enforcement for Compliance with Phase II Permits The main focus for enforcing compliance with Phase II permits will be to incorporate the long-term CSO control plan through a civil judicial action, administrative order, or other enforceable mechanism requiring compliance with the CWA and imposing a compliance schedule with appropriate milestone dates necessary to implement the plan.

In general, a judicial order is the appropriate mechanism for incorporating the above provisions for Phase II. Administrative orders, however, may be appropriate for permittees whose long-term control plans will take less than five years to complete and for minors that have complied with the final date of the enforceable order for compliance with their Phase I permit. If necessary, any of the nine minimum controls that have not been implemented by this time should be included in the terms of the judicial order.

D. Penalties

EPA is prepared not to seek civil penalties for past CSO violations if permittees have no discharges during dry weather and meet the objectives and schedules of this Policy. Notwithstanding this, where a permittee has other significant CWA violations for which EPA or the State is taking judicial action, penalties may be considered as part of that action for the following:

1. CSOs during dry weather;
2. violations of CSO-related requirements in NPDES permits, consent decrees, or court orders which predate this policy; or
3. other CWA violations.

EPA will not seek penalties for past CSO violations from permittees that fully comply with the Phase I permit or enforceable order requiring compliance with the Phase I permit. For permittees that fail to comply, EPA will exercise its enforcement discretion in determining whether to seek penalties for the time period for which the compliance schedule was violated. If the milestone dates of the enforceable schedule are not achieved and penalties are sought, penalties should be calculated from the last milestone date that was met.

At the time of the judicial settlement imposing a compliance schedule implementing the Phase II permit requirements, EPA will not seek penalties for past

CSO violations from permittees that fully comply with the enforceable order requiring compliance with the Phase I permit and if the terms of the judicial order are expeditiously agreed to on consent. However, stipulated penalties for violation of the judicial order generally should be included in the order, consistent with existing Agency policies. Additional guidance on stipulated penalties concerning long-term CSO control and attainment of WQS will be issued.

(Approved by the Office of Management and Budget under Control Number 2040-0170)

Glossary

Acute toxicity Severe (usually fatal) toxic effects that rapidly affect organisms due to chemical exposure.

Advanced treatment Treatment applied for a high degree of pollutant removal in which a significant amount of colloidal and dissolved solids are removed. The process chain generally includes preliminary treatment, secondary biological treatment, and chemical clarification.

Advective transport Physical transport of water and associated concentrations from higher to lower hydraulic potential, exclusive of dispersion/mixing.

Aesthetics Of or pertaining to the sense of attractiveness.

Algorithm A calculation procedure.

Aperture size Size of opening between screen material.

Assimilation In water bodies, the process that removes pollutants and/or their impacts.

BAT Best available technology.

BCT Best conventional pollutant technology.

Best management practices (BMT) Non- and low-structurally intensive storm and CSO pollution control methods that have multiple benefits, including drainage network enhancement and flood control; groundwater recharge; aesthetic enhancement; pavement cleansing and reduction of dust, dirt, litter, and debris; subpotable water reuse; and so on.

Bioaccumulation Accumulation of trace substances in the tissues of biological organisms.

Biochemical oxygen demand (BOD) An operational measure of potential for depletion of dissolved oxygen by the biological and chemical degradation of organic material by bacteria.

Biologic diversity A measure of the variety of biological organisms.

Biological treatment processes Means of treatment in which bacterial or biochemical action is intensified to stabilize, oxidize, and nitrify the unstable organic matter present. Trickling filters, activated sludge processes, and lagoons are examples.

BMP Best management practices.

BOD Biochemical oxygen demand.

BPJ Best professional judgment.

BPT Best practicable control technology.

Calibration See *validation*.

Catch basin A chamber or well, usually at the street curbline, for the admission of surface water to a sewer or subdrain, having at its base a sediment sump to retain grit and detritus below the point of overflow; a stormwater inlet, by contrast, does not have a sump and does not trap sediment.

Catchment The area producing the runoff passing a particular channel or stream location.

Chlorophyll a A green pigment found in photosynthetic organisms; used as an indicator of algal biomass.

Chronic toxicity Toxic effects from long-term exposure to chemicals; usually nonfatal to organisms. Effects often consist of reduced growth and reproduction, as well as other physiological impacts.

Collection system control A method of abating storm-generated or CSO pollution in the collection or drainage system.

Combined sewer A sewer receiving intercepted surface (dry- and wet-weather) runoff, municipal (sanitary and industrial) sewage, and subsurface waters from infiltration.

Combined sewer overflow (CSO) Flow from an outfall (discharge conduit) of a combined sewer collection system in excess of the interceptor capacity that is discharged into a receiving-water and/or an auxiliary CSO control (storage) treatment system.

Composites Water samples comprised of a series of equal volume samples taken over interval time periods *(temporal composite)* or spatial locations *(spatial composite)*.

Computer model A model in which the mathematical operations are carried out on a computer.

Continuous simulation The use of a model to simulate the response of a catchment to a series of storm events and the hydrologic processes that occur between them.

Cost-benefit relationship The relationship between unit costs and unit benefits, usually represented as a curve.

Cost-effective solution A solution to a problem that has been identified as financially optional (e.g., the solution associated with the knee-of-the-curve of a cost-benefit relationship).

Critical design conditions Environmental and flow conditions chosen to represent the conditions under which compliance with water-quality standards, criteria, or objectives is desired.

CSO Combined sewer overflow.

DAF Dissolved air flotation.

Desktop analytical models Mathematical models comprised of closed-form solutions that can be readily calculated using a handheld calculator or computer.

Deterministic mathematical model A mathematical model designed to produce system responses or outputs to temporal and spatial stimuli or inputs.

Detritus A combination of relatively heavy, coarse solids comprised of sand, grit, and organic matter usually having a Stokes's-law settling velocity equal to or greater than a particle with a 0.2 mm effective diameter and a 2.65 specific gravity.

Detention The slowing, dampening, or attenuating of flows either entering the sewer system or within the sewer system by temporarily holding the water on a surface area, in a storage basin, or within the sewer itself.

Detention time The time period that flow is detained in a storage/sedimentation basin or tank.

Digital computer model A computer model constructed and operated in the binary representations of a digital computer.

Disinfection The killing or inactivation of human-disease-causing microorganisms or pathogens.

Dispersion Pollutant or concentration mixing due to turbulent physical processes.

Dissolved oxygen deficit Difference between saturated dissolved oxygen and ambient concentrations.

Distributed model A model in which the physical heterogeneities of the catchment are included.

Diurnal Occurring daily, or over the duration of a day.

DNMRF Dual-media high-rate filtration.

Domestic sewage Sewage derived principally from dwellings, business buildings, institutions, and the like. It may or may not contain groundwater.

Drogue A floating marker that drifts with the prevailing surface current.

Dry-weather flow (DWF) Usually referred to as the flow in a combined sewer system without stormwater.

DWF Dry-weather flow.

Dual treatment Those processes or facilities designed for operating on both dry- and wet-weather flows.

Dynamic flow routing The movement of water in a channel or over a surface that is modeled with both the continuity of mass and momentum equations. This allows the simulation of the propagation of waves (or disturbances) in the upstream and downstream directions.

Ecological habitat The environmental niche in which an organism lives.

End-of-pipe impacts Impacts that occur in the immediate vicinity of an outfall.

ENR *Engineering News Record.* A recognized magazine providing accepted construction-cost indices.

Enteric Relating to the intestinal tract.

Eutrophication The process of aging whereby the increase of mineral and organic nutrients favors aquatic plants over animal life and results in increasing daily variations in dissolved oxygen concentrations, reduced biologic diversity, and reduced water clarity.

Exceedence interval The average interval of time between values that exceed a specified magnitude.

Execution time See *run time.*

FC Fecal coliform.

First flush The condition, often occurring in storm-sewer discharges and CSOs, in which a disproportionately high pollutional load is carried to the first portion of the discharge or overflow.

Floatables Large floating material sometimes characteristic of sanitary wastewater and storm runoff.

Flow routing The mathematical modeling of flow or flood waves through a channel (e.g., sewer) or over a surface.

Flux The flowrate per unit cross-sectional area.

Foul sewer The connecting sewer between the swirl or helical bend regulator/concentrator and the interceptor that provides both a pathway to the interceptor for sanitary sewage flow during dryweather flow and concentrated combined sewage during stormflow periods.

Frequency of return The rate at which a particular type storm can be expected to occur (e.g., one year); such storms are classified by storm intensity and duration.

FS Fecal streptococcus.

Helical bend regulator/concentrator A CSO control device having an arc or bend shape, providing the same dual functions as cited for the *swirl regulator/concentrator.*

HGMS High-gradient magnetic separation or separator.

Hydraulic loading The flowrate per unit plan or surface area in sedimentation facilities or cross-sectional area in screening and filtration facilities (see also *flux*).

Hydrodynamic Nonsteady-state (dynamic) hydraulic conditions.

Incremental impacts Fractional portions of an overall impact attributable to different individual sources.

Infiltration The process whereby water enters a sewer system and service connections underground through such means as, but not limited to, defective pipes, pipe joints, connections, or manhole walls. Infiltration does not include, and is distinguished from, inflow.

Inline storage A type of storage that has no pumping requirements and can consist of either storage within the sewer (in-sewer/in-pipe) or channel, or storage in the inline basins.

In-pipe storage See *inline storage.*

Interceptor A sewer designed to receive dry-weather flow from a number of transverse combined sewer trunks and additional quantities of intercepted surface runoff during low-flowing sanitary sewage periods and to convey such waters to a point for treatment.

Intertidal Related to events during a tidal cycle.

Intratidal Related to events encompassing a full tidal cycle, or averaged for a tidal cycle.

Kinematic flow routing The movement of water in a channel or over a surface that is modeled with the continuity of mass equation and a simple relationship between flow and depth (replacing the momentum equation). This allows only the simulation of the propagation of waves (or disturbances) in the downstream direction.

Knee-of-the-curve The point along a cost-benefit curve at which there is a noticeable change in the quantity of cost for an increment of benefit.

Lagrangian framework A mathematical modeling framework based on a moving reference point in contrast to a fixed coordinate reference point (named after the French mathematician J. L. Lagrange).

Lateral A sewer that has no other common sewer discharging into it.

Longitudinal dispersion One-dimensional dispersion (mixing) occurring along the length of the stream or estuary.

Lumped model A model in which the physical characteristics of the catchment are assumed to be homogeneous.

Mass conservation Scientific principle that matter is neither created nor destroyed in nonradioactive systems.

Mass yield The mass of a pollutant that reaches the downstream point of concern over a certain time period.

Mathematical modeling Application of mathematical formulae to represent the processes and effects of natural and synthetic systems for the purpose of forecasting responses to different conditions and inputs.

Matrix A fibrous ferromagnetic material in the canister of the high-gradient magnetic separator treatment system.

Model Any representation of a system by something other than the system itself.

Model calibration Refinement of mathematical model parameters and coefficients through comparison to data by making scientifically consistent and rational adjustments.

Model parameter A quantity that cannot vary in a particular model run.

Model variable A quantity that can vary in a particular model run.

Monte Carlo probabilistic simulation A statistical modeling approach used to assess the expected frequency and magnitude of water-quality impacts by running repetitive simulations using statistically selected inputs for rainfall, pollutants, flows, temperature, and so on.

Multiple-event simulation The use of a model to simulate the response of a catchment to several independent precipitation events (typically a number of different design storms).

Nitrification Biological oxidation of nitrogen compounds by nitrobacteria that consume dissolved oxygen.

Nonpoint Diffuse; not attributable to a particular location.

Nonpoint-source pollution Any unconfined and nondiscrete conveyance from which pollutants are discharged.

Off-line storage A type of storage that requires detention facilities (basins or tunnels) and facilities for pumping storm flow either into or out of the detention facilities.

Parameter optimization See *validation.*

Pathogen A disease-causing microorganism.

Pathogenic bacteria and viruses Bacteria and viruses capable of causing disease in humans.

Peak-and-shape parameter A model parameter that strongly influences the peak, rising, and receding legs of a runoff hydrograph.

Photosynthesis The process whereby green plants (including algae) convert incident light to chemical energy and synthesize carbohydrates from carbon dioxide and water with the simultaneous release of oxygen.

Physical treatment processes Means of treatment in which the application of physical forces predominate. Screening, sedimentation, flotation, and filtration are examples. Physical treatment operations may or may not include chemical additions.

Physical with/without chemical treatment processes Means of treatment in which the removal of pollutants is brought about primarily by physical processes, with or without chemical addition to enhance removal efficiency.

Production run A run of a validated model intended to produce predictive information.

PVC Polyvinyl chloride.

RBC Rotating biological contactor.

Reaeration rate Rate of mechanical replenishment of dissolved oxygen from the atmosphere via wind, waves, and turbulent mixing.

Receiving waters Natural or synthetic water systems into which materials are discharged.

Recurrence interval See *frequency of return.*

Recurrent frequency The historical frequency at which a condition or situation occurs.

Regulator A structure that controls the amount of sewage entering an interceptor by storing in the upstream trunk line or by diverting some portion of the flow to an outfall.

Run A single execution of a model.

Run time The time needed to execute a model run on a computer.

Saline intrusion The intruding of saline-rich seawater (which is more dense) into a predominantly freshwater system.

Sanitary sewer A sewer that carries liquid and water-carried wastes from residences, commercial buildings, industrial plants, and institutions, together with relatively low quantities of ground-, storm-, and surface waters that are not admitted intentionally.

Sanitary wastewaters Wastewater of human origin.

Satellite facilities CSO storage/treatment facilities at remote locations upstream of the dry-weather flow sewage treatment plant and usually at the regulator/overflow site.

Sediment oxygen demand Biochemical consumption of dissolved oxygen in overlying waters by decaying sediments across the water-sediment surface.

Sensitivity analysis The variation of model parameters to determine the sensitivity of the model to each parameter.

Sewer A pipe or conduit generally closed, and normally not flowing full, for carrying sewage or other waste liquids.

Sewer flushing Flushing applied to combined sewer systems during dry-weather flow periods to remove settled material periodically or as it accumulates and to hydraulically convey it to the treatment facilities, thus preventing resuspension and overflow of a portion of the solids during storm-flow periods and lessening the need for CSO treatment.

Sewerage System of piping, with appurtenances, for collecting and conveying wastewaters from source to treatment and/or discharge.

S.G. Specific gravity.

Simulation The application of a model.

Simulation time The time simulated by a model run.

Single-event simulation The use of a model to simulate the response of a catchment to a single precipitation event (typically a design storm).

Source control A method of abating storm-generated or CSO pollution at the upstream, upland source where the pollutants originate and/or accumulate.

Spatial discretization The level of detail in describing the changes in space of a system, such as a catchment or sewer system, for modeling purposes.

Saint-Venant equations The equations of continuity of mass and momentum used to model flow in a channel or over a surface.

SS Suspended solids.

Steady state In a state of steady, unchanging conditions.

Stochastic mathematical model A mathematical model designed to produce a series of random responses (e.g., runoff) that have the statistical characteristics of historical responses of the system.

Storage The slowing, dampening, or attenuating of storm-generated or combined-sewer flows either entering the sewer/drainage system or within the sewer/drainage system by temporarily holding the flow on a surface area, in a storage basin, within the sewer itself, or within a receiving water.

Storm duration The period over which rainfall occurs.

Storm flow Overland flow, sewer flow, or receiving-stream flow caused totally or partially by surface storm runoff, storm-related subsurface infiltration, or snowmelt.

Storm frequency See *frequency of return.*

Storm intensity The rate of rainfall usually expressed as inches per hour.

Storm sewer A sewer that carries intercepted surface runoff, street wash, and other wash waters, or drainage, but excludes domestic sewage and industrial wastes except for unauthorized cross-connections.

Storm-sewer discharge Flow from a storm sewer that is discharged into a receiving water or a sewer system.

Stormwater Water resulting from precipitation that percolates into the soil; runs off freely from the surface; or is captured by storm sewer/drainage, combined sewer, and, to a limited degree, sanitary sewer facilities.

Stratification Layering of water caused by differences in density due to temperature and dissolved solids.

Streeter-Phelps equation Closed-form analytical mathematical solution for steady-state calculation of a dissolved oxygen in nondispersive stream systems.

Subcatchment A portion of a catchment producing the runoff that passes a channel or stream location upstream of the location defining the catchment.

Supernatant The relatively clear liquid layer above the sediment layer in the vertical column.

Superposition concept Mathematical approach whereby a complex process is represented by the summation of its independent parts.

Surcharging The transition between open channel flow and pressure flow in sewers.

Surface runoff Precipitation and water (e.g., street wash) that falls onto the surfaces of roofs, streets, ground, and so on, and is not absorbed or retained by that surface, thereby collecting and running off.

Swirl regulator/concentrator A cylindrically-shaped CSO control device that provides the dual function of a regulator and a solids-liquid concentrator. As a concentrator, it achieves good removal of the heavier settleable solids fraction in CSO. (See also *regulator.*)

System A network of interacting components capable of responding to one or more stimuli (e.g., watershed, combined sewer system, receiving water).

TC Total coliform.

Temporal discretization The level of detail in describing the changes over time in a system such as a catchment or sewer system for modeling purposes.

Time step The increment of time used to discretely introduce changes in the system.

Total solids The entire quantity of solids in the liquid flow or volume including the dissolved and particulate fractions.

Toxicity The degree to which a pollutant causes physiological harm to the health of an organism.

Trace metals Metals present in small concentrations. From a regulatory standpoint, this usually refers to metal concentrations that can cause toxicity at trace concentrations.

Transient Temporary; of short duration.

Trunk A sewer, also known as a main sewer, that receives the discharge of one or more submain sewers.

Urban runoff Surface runoff from an urban drainage area that reaches a stream or other body of water or a sewer/channel.

Validation The adjustment of model parameters so that the model output adequately mimics the system being modeled.

Verification The testing of a model to determine if it is functioning properly.

Volume parameters A model parameter that strongly influences the volume of runoff produced from a precipitation event.

Water-quality criteria A threshold value or concentration for a pollutant or pollutant effect as chosen by regulatory agencies to distinguish between acceptable and nonacceptable environmental conditions; usually chosen based on laboratory observations of organism response.

Water-quality standard A threshold value or concentration enforced by law as a requirement to maintain acceptable environmental water-quality conditions; usually chosen based on laboratory observations of organism response.

Wet-weather flow Usually referred to as the flow in a combined sewer system with stormwater but may also constitute the flow in a separate storm or sanitary drainage system with stormwater.

Index